Student Book

NEW GCSE MATHS
Functional Skills

Andrew Bennington • Andrew Manning • Dr Naomi Norman

William Collins' dream of knowledge for all began with the publication of his first book in 1819. A self-educated mill worker, he not only enriched millions of lives, but also founded a flourishing publishing house. Today, staying true to this spirit, Collins books are packed with inspiration, innovation and practical expertise. They place you at the centre of a world of possibility and give you exactly what you need to explore it.

Collins. Do more.

Published by Collins
An imprint of HarperCollinsPublishers
77 – 85 Fulham Palace Road
Hammersmith
London
W6 8JB

Browse the complete Collins catalogue at
www.collinseducation.com

10 9 8 7 6 5 4 3 2 1

ISBN-13 978-0-00-741006-4

Andrew Bennington, Andrew Manning and Dr Naomi Norman assert their moral rights to be identified as the authors of this work

British Library Cataloguing in Publication Data

A Catalogue record for this publication is available from the British Library

Commissioned by Katie Sergeant
Project managed by Aimee Walker and Sue Chapple
Edited and proofread by Joan Miller, Christine Vaughan and Rosie Parrish
Cover design by Julie Martin
Cover photography by Caroline Green
Illustrations by Ann Paganuzzi
Design and typesetting by Linda Miles, Lodestone Publishing Limited
Production by Kerry Howie
Printed and bound by L.E.G.O. S.p.A. Italy

With thanks to Dr Naomi Norman, Howard Marsh and Jonathan Miller for their support and contribution of ideas.

Contents

Introduction

Collins Functional Skills Student Book enables you to build, apply and strengthen functional and process skills and mathematics in interesting and real-life scenarios. These skills are an integral part of the new GCSE specifications and level 2 functional skills.

What this book is about

- Showing you that mathematics can be used to tackle real-life problems.
- Helping to improve your confidence in using mathematics.
- Encouraging you to apply the mathematical skills you have in new situations.
- Developing your ability to interpret results and decide how successful you have been.

What you will get out of it

- The chance to use the mathematical skills you have in a realistic context.
- The opportunity to work with others and talk about mathematics.
- An understanding of why mathematics can be useful in life and work.
- Greater confidence about your own ability to use mathematics.
- Interesting and enjoyable mathematics lessons.

Some things to think about

- People who can use mathematics confidently get the most out of life and work. That could be you.
- Do not worry about making mistakes. A lot of the tasks in this book have more than one right answer. We want to know **your** solution and we want you to be able to explain how you obtained it.
- Mathematics can be interesting, useful and exciting. We hope to convince you that this is true.

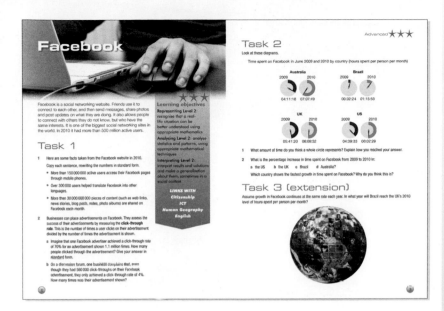

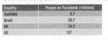

Star ratings

Tackle challenges at different levels, including beginner, improver and advanced, to help build your confidence in using maths in real life.

Learning objectives

Understand the skills you will be practising in each activity and how they match up with the Functional Skills Standards.

Cross-curricular links

Identify how the maths in each topic links with other subjects.

Setting the scene

Think about the context of each topic by reading the interesting introductions.

Extension tasks

Give yourself more of a challenge with the extension tasks.

Self-assessment

Think about and discuss how you found the tasks in each topic.

Birds

The Royal Society for the Protection of Birds (RSPB) and The Wildfowl & Wetlands Trust (WWT) are two of the largest UK bird charities. They both protect and care for birds and make sure that the UK offers good habitats for birds to live in. They also track bird populations.

Task 1

Volunteers for a bird charity often conduct bird surveys.
What would be the more suitable kind of data collection for each of the following surveys? Should they use **sampling** or **observation**?

1   The number of different kinds of bird seen in a garden one weekend.

2   The number of responses to a recording of an owl's hoot through a loudspeaker late one spring night.

3   The population of nightingales migrating to the UK in spring.

4   The number of birds nesting in and around a local park.

5   The number of days it takes birds to eat a bag of bird food hung in a school playground.

6   The frequency of cuckoo calls in a woodland during a weekend in April.

7   The number of days it takes for a swan's egg to hatch.

8   The typical number of birds in a flock of sparrows.

Task 2

Eight children recorded bird sightings in their gardens for an hour, early one Sunday morning in February 2010.

The table lists the top 10 birds, and the numbers the children spotted.

	Bird species	Toby	Edward	Isy	Louise	Ross	Miguel	Sophie	Jamie
1	House sparrow	5	3	2	4	5	1	5	4
2	Starling	3	0	5	1	3	6	2	4
3	Blackbird	2	4	5	2	0	4	3	1
4	Chaffinch	1	4	0	6	3	1	2	2
5	Blue tit	3	6	1	4	2	2	0	0
6	Wood pigeon	4	3	3	0	2	3	1	2
7	Robin	1	2	3	3	2	2	1	2
8	Collared dove	2	0	0	2	0	0	2	0
9	Goldfinch	0	0	0	0	3	1	0	0
10	Great tit	0	1	0	0	0	0	0	1

1 What is the modal number of:

    a house sparrows   **b** blue tits?

2 What is the median number of:

    a chaffinches   **b** collared doves?

3 What is the mean number of:

    a starlings   **b** robins

    c wood pigeons   **d** collared doves?

Round your answers to **3c** and **3d** to the nearest whole numbers.

4 Find the range for sightings of:

    a wood pigeons   **b** robins.

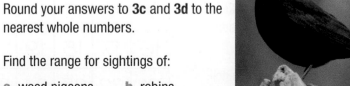

Task 3 (extension)

Here are the top 10 results for a bird charity's 2010 garden birdwatch survey. It was conducted by 729 people for an hour, again over one weekend in February 2010.

Bird species	Average number per garden	Bird species	Average number per garden
House sparrow	3.88	Wood pigeon	1.89
Blackbird	3.23	Robin	1.59
Starling	3.27	Great tit	1.42
Blue tit	2.48	Collared dove	1.34
Chaffinch	2.13	Goldfinch	1.12

1   What average do you think the charity used – mode, median or mean? Explain your answer.

2   Write two sentences comparing the results for starlings and robins for the charity's survey and the children's survey.

3   Round the charity's averages for starlings and robins to the nearest whole number. Now again write two sentences comparing the results for these birds for the charity's survey and the children's survey.

4   Whose survey results are likely to be more accurate – the children's or the charity's? Explain your answer.

Task 4

The children represented their surveys in different ways, but forgot to write their names on their diagrams. Look at the data sheet and then use Task 2 to help you decide whose name to write on each diagram.

Task 5

The children decided to conduct another bird survey. This time they were looking for birds of prey within a large area of open countryside. They chose an area between four local villages and linked them, to make a square. Then they realised they could not cover the whole area. Therefore, they drew a grid and gave each cell in the grid a number.

1	2	3	4	5
6	7	8	9	10
11	12	13	14	15
16	17	18	19	20
21	22	23	24	25

They wrote each number on a piece of paper and put them in a pencil case. Then they each randomly picked out one piece of paper. This was the number of the cell they would cover for the survey.

21	2	10	19	24	11	13	3
Toby	Edward	Isy	Louise	Ross	Miguel	Sophie	Jamie

The children displayed their survey results like this.

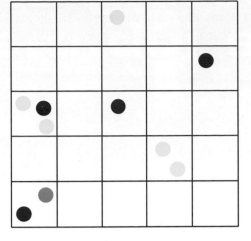

Sparrowhawk ●

Kestrel ●

Buzzard ●

Red kite ●

Peregrine falcon ●

1   Which bird was spotted:

    a most regularly   **b** not at all?

2   Which of the children:

    a did not spot any birds of prey

    b spotted only one bird of prey

    c spotted two birds of prey

    d spotted three birds of prey?

3   What is the mean number of birds spotted in a chosen cell?

4   The children used the following calculation to estimate the number of birds of prey in the total area:

    mean number of birds for the chosen cells × total number of cells

What answer did the children get?

5   Round the children's answer to the nearest whole number.

Task 6

Use ideas from Tasks 1–5 to plan and conduct your own bird survey. You may work in small groups or as a class.

HOW DID YOU FIND THESE TASKS?

- What did you find easy or difficult about these tasks?
- Did you work on your own, in pairs or in groups, and how did this help or hinder your approach and success with these tasks?
- What did you learn about how maths is used and applied in real-world situations?

Getting ahead in the job market

No one wants to be unemployed but it's not always easy to find a job.

How can you improve your chances? Aim to achieve your full potential by getting good grades and the best qualifications you can. After GCSEs, you might start an apprenticeship, learning and earning while you work, or you might go to sixth-form college to study A-levels and on to university.

Task 1

There are many different ways to measure unemployment. Unemployment figures for April 2010 were about 4.8%.

1   Look carefully at this graph. Why might it be misleading?

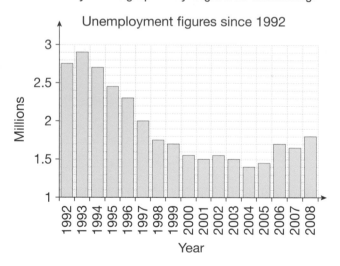

Unemployment figures since 1992

2   Put the information from the block graph above into a line graph. Include the data for April 2010.

3   If you were going to publish this data in a report on unemployment, which graph would you use? Give reasons.

Unemployment reaches 2.46 million in April 2010

RADIO 4

4   Look at the maps, below. How has the unemployment rate changed?

Percentage of people claiming unemployment benefits

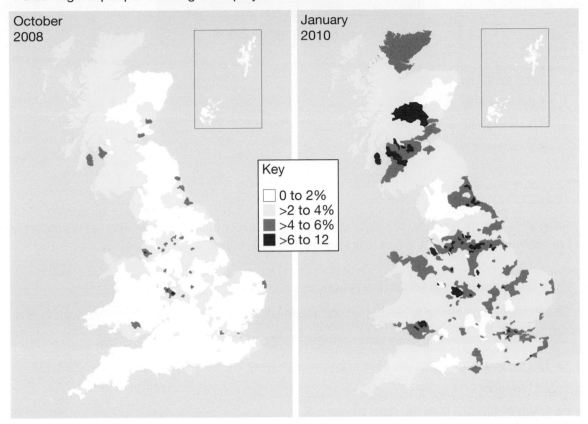

October 2008

January 2010

Key

☐ 0 to 2%
  >2 to 4%
  >4 to 6%
  >6 to 12

Task 2

Gaining higher qualifications, such as A-levels or a degree, can mean that you might get a better-paid job when you start work.

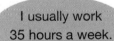

1   The Government sets national minimum wage figures that employers have to follow. Here are the figures for September 2010.

Age (years)	Hourly rate
Under 18	£3.57
18–21	£4.83
22 or more	£5.80

Most people work a 35-hour week and are paid for 52 weeks in the year.

a Calculate the annual gross pay for:
- a school-leaver aged 16
- a 19-year-old
- a 25-year-old worker.

b Your **net pay** is the amount you are left with, after all deductions have been made. Deductions include tax and national insurance.

> Tax allowance of £6475 then 20% tax
> National insurance at 6%

Calculate the net pay of:

- a 16-year-old
- a 19-year-old
- a 25-year-old.

Task 3

You will want to look for a job that will pay you enough to cover your living expenses. This will depend on where you live and who you live with and how many bills you will have to pay.

1   Your living expenses will fall into two categories.

Essentials include paying to keep a roof over your head, food on the table and any other monthly bills.

Luxury items include take-away meals and visits to the cinema and even new clothes.

a List the **essential items** you need to pay for each week and the **luxury items** you would consider purchasing.

b Estimate the cost of all the items on your list.

2   Imagine you were an 18-year-old earning the national minimum wage.

a Work out your salary, after deductions.

b How much money would you have left over, after paying for your essentials?

c Would you have enough money for your luxury items?

Every year, there is a rise in the cost of living. This means that the cost of buying things increases. This is usually about 3%.

d Calculate the cost of the items on your new list after a 3% rise in the cost of living. Can you still afford everything on your list?

e Find advertisements for jobs that you think you could apply for, in your area. Write down the hourly rate or salary. Compare the jobs you have found and find the average wage. Could you afford the essential and luxury items on your list if you took one of the jobs in your area? Show your working.

Task 4 (extension)

There are many different ways to pay for goods.

1   Complete the table on the data sheet.

2   Design a data-collection sheet and use it to find out other people's opinions of the methods of payment listed in the table.

3   Present your findings in an appropriate graph, chart or table.

4   Comment on your results.

HOW DID YOU FIND THESE TASKS?

- What did you find easy or difficult about these tasks?
- Did you work on your own, in pairs or in groups, and how did this help or hinder your approach and success with these tasks?
- What did you learn about how maths is used and applied in real-world situations?

Blood donors

A **blood donor** is someone who gives some of their own blood to help others who may need it.

Blood that is donated may be given to someone who has lost blood because they have had an accident, or to someone whose own blood does not work properly because they are ill. A person who receives blood is called a **recipient**.

A **blood transfusion** is the process of transferring blood from the donor to the recipient.

Task 1

Anyone who is aged 17 or older, and weighs over 7 st 12 lb, can volunteer to donate blood.

1   How many pounds (lb) are there in 7 st?

2   How many pounds are there in 7 st 12 lb?

3   How many grams is this?

4   How many kilograms is this?

| 1 lb ≈ 450 g |

5   Here are the weights and ages of students in a Year 12 class.
Which four students would not be allowed to give blood?
Give reasons for your answers.

Name	Age	Weight (kg)
Fred	17	52.4
Coral	17	50.2
Richard	17	55
Lucy	17	49.9
Emily	17	49.6
Abbie	17	49.06

Name	Age	Weight (kg)
Phoebe	16	49.6
Ginny	17	50.3
Jordan	17	59.7
Georgia	17	50
Charlie	17	44.5
Phil	17	57

Name	Age	Weight (kg)
Heidi	17	49.9
James	17	54.2
Carla	17	49.56
Caroline	17	50.7
Greg	17	60.5
Roman	16	50

Task 2

Everyone's blood belongs to one of four groups: A, B, AB or O.

Here are the blood groups of the students in the Year 12 class.

Name	Blood group	Name	Blood group	Name	Blood group
Fred	A	Phoebe	O	Heidi	A
Coral	O	Ginny	A	James	O
Richard	O	Jordan	O	Carla	B
Lucy	A	Georgia	B	Caroline	O
Emily	O	Charlie	O	Greg	A
Abbie	A	Phil	AB	Roman	O

1   Construct a frequency table to record the blood groups of all the Year 12 students.

2   Which blood group is most common among this group of students?

3   Which blood group is least common among this group of students?

4   How many students are there in the Year 12 class?

5   Use your frequency table to work out the probability of choosing a student from this Year 12 class whose blood group was:

a AB   **b** A   **c** O   **d** B.

Give your answers as fractions in their lowest terms.

Task 3 (extension)

Compare the blood groups of students in the Year 12 class (from Task 2) to those of the UK population.

Blood group	Percentage of the UK population
O	44%
A	42%
B	10%
AB	4%

Write three or four sentences describing your comparison.

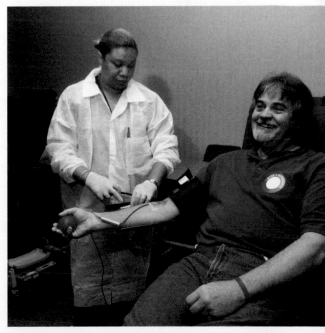

Task 4

Not everyone can donate or receive blood from everyone else. It works like this.

Blood group	A	B	AB	O
Can donate to	A	B	AB	A
	AB	AB		B
				AB
				O
Can receive from	O	O	A	O
	A	B	B	
			AB	
			O	

Use this table and the table from Task 2 to complete this task.

1 **a** List all the Year 12 students who can donate blood to group B.

(Remember that four students are not allowed to donate blood.)

b List all the Year 12 students who can receive blood from group A.

(Remember that all students are allowed to receive blood.)

2 Copy and complete the sentence below. Replace the first blank with:

donate blood to   or   receive blood from

Replace the second blank with one of the following groups.

A   B   AB   O

Do this four times to produce four questions.

How many of the Year 12 students can _____ group ___?

Now swap with a partner and answer each other's four questions.

Remember that four students are not allowed to donate blood, although they can receive blood.

Task 5 (extension)

A boy and a girl are in a traffic accident and both need a blood transfusion.
The boy's blood is group O and the girl's is group AB.
What is the probability of each of them finding a donor among the
Year 12 students?
Which of them has the better chance?

Task 6

The Year 12 class divides into two groups, donors and recipients, as follows.

Donors		Recipients	
Name	**Blood group**	**Name**	**Blood group**
Greg	A	Abbie	A
Carla	B	Heidi	A
Ginny	A	Roman	O
Caroline	O	James	O
Jordan	O	Charlie	O
Coral	O	Lucy	A
Emily	O	Phil	AB
Richard	O	Phoebe	O
Fred	A	Georgia	B

Try different ways of matching donors to recipients.

Remember, not everyone can donate or receive blood from everyone else.

1 Are there any students who can donate to everyone?

2 Are there any students left over, who need a donor but who do not have one?

3 Caroline and Phil decide to swap places. Try different ways of matching donors to recipients now. What do you notice?

Task 7

Imagine you work for the Blood Donor Service. Use what you have learnt in Tasks 1–6 above to write a leaflet encouraging people to give blood.

HOW DID YOU FIND THESE TASKS?

- What did you find easy or difficult about these tasks?
- Did you work on your own, in pairs or in groups, and how did this help or hinder your approach and success with these tasks?
- What did you learn about how maths is used and applied in real-world situations?

Tuning in

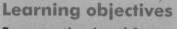

The way we listen to the radio has changed significantly in recent years. New digital stations are added to the list regularly, although some do not last for long. These can be accessed through digital radios, through televisions with digital reception (satellite, cable or freeview) or mobile phones.

Many stations are also available over the internet. This means we can listen to local radio even when we are away from home. Many broadcasters also allow us to use their websites to listen to programmes a week or more after the original date of transmission.

Learning objectives

Representing Level 1: understand practical problems and use scales

Analysing Level 1: perform calculations with large numbers; calculate percentages

Interpreting Level 1: draw and interpret graphs

LINKS WITH
Music
Science
Media studies

Task 1

Here is a list of the frequencies of some national radio stations.

Radio station	Frequency (FM)
Station A	97.6–99.8
Station B	88.1–90.2
Station C	90.2–92.4
Station D	92.4–96.1 103.5–104.9
Station E	99.9–101.9

1 Copy the scale below and mark the radio frequencies on it.

Frequency	88	90	92	94	96	98	100	102	104	106	108
Station											

2 What radio stations can you receive in your area? You could find out from the internet or a local newspaper. Mark the frequencies on the scale.

Task 2

The graph on the right shows the number of people who listen to each of 10 national radio stations every week.

1 Which station is the most popular?

2 How many listen to that station every week?

Station A has an audience share of 9.3%. This means that of all the people who could listen to this station, 9.3% actually do.

3 A radio station in Hampshire reports that it has 12.6% of the audience share, but it is not one of the top 10 radio stations listed in the graph. How can this be true?

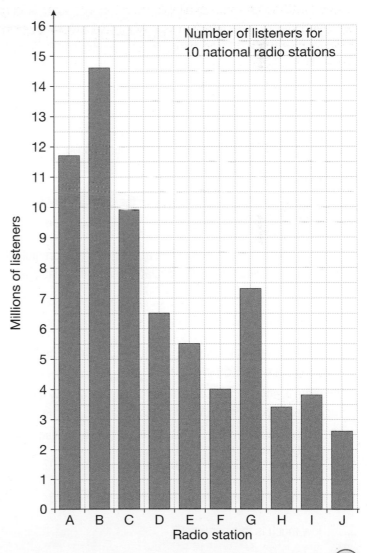

Number of listeners for 10 national radio stations

Task 3

Research has shown that the number of people who listen to the radio over the internet is growing rapidly. Read the information below, which was gathered in 2010.

- 16.1 million people use the internet to listen to radio.

- 15 million listen live.

- 13.5 million use the internet to listen to programmes after they have been transmitted live.

- 11.6 million people use the internet to listen to radio at least once a week.

- The average user of repeat radio transmissions listens to 1.65 programmes each week.

- Three-quarters of repeat transmission listeners listen to the same amount of live radio as they did before.

- Almost half of repeat transmission listeners now listen to radio programmes that they did not previously listen to.

Study the following table.

Internet radio listeners over a seven-year period							
Year	2002	2003	2004	2005	2006	2007	2008
Percentage of adults who listen regularly to internet radio	2.7%	3.6%	7.3%	9.3%	11.9%	13.5%	14.0%

1   Draw a line graph to show this information.

2   According to the information in the points above, how many people used the internet to listen to the radio at least once a week in 2010?

3   The potential listening audience in the UK is about 50 million. What percentage of these regularly listened to the radio (at least once a week) on the internet in 2010?

4   Your graph stops at 2008. Look at the graph and predict what might happen in 2009 and 2010. Does the 2010 figure you have just calculated seem reasonable from your graph?

Task 4

In the UK, BBC national radio receives its income from the television licence fee we all have to pay. The annual licence cost was £145.50 in 2010.

Most of the licence income goes towards paying for new television programmes. Only about £2.68 of each licence fee payment goes towards radio (including internet radio).

In 2010, the income from the television licence fee totalled £3450 million.

1   How many television licences were bought in 2010?

2   How much of the licensing fee was spent on radio?

3   How much is this per week?

Task 5

Commercial radio stations get their money from advertising.

The cost of an advertisement depends on its length and how many people listen to the radio station.

A local commercial radio station has an average audience of 20 000. A 30-second radio advertisement costs £2 for every 1000 listeners.

1   How much would a 30-second advertisement cost on this station?

2   Imagine that this radio station broadcasts eight advertisements per hour. How much money will this radio station receive per week?

Task 6 (extension)

Conduct a survey among your age group to see whether they listen to the radio and, if they do, which radio stations they listen to. You will need to construct a questionnaire with appropriate questions so that you can gather the information to complete this task.

1   Which is the most popular radio station?

2   What percentage of the people surveyed listen to the most popular station?

3   What is the average number of hours per day people listen to the radio?

HOW DID YOU FIND THESE TASKS?

- What did you find easy or difficult about these tasks?
- Did you work on your own, in pairs or in groups, and how did this help or hinder your approach and success with these tasks?
- What did you learn about how maths is used and applied in real-world situations?

Making your own fuel

As fuel prices keep rising, people are starting to think seriously about using vegetable oil to run cars and other motors. In fact, older diesel engines will run quite happily – although a bit smokily – on standard vegetable oil from the supermarket. There are even companies that use old oil from fish and chip shops to produce fuel. Some people have commented that prices of vegetable oil in supermarkets are reflecting this, as prices show an upward trend.

Fuels from vegetable sources are known as biofuel, or **biodiesel fuel**; rape seed is now being grown commercially in the UK to provide biodiesel fuel.

Task 1

Most new projects involve start-up costs, and making fuel from vegetable oil is no different. Look at the data sheet to find a list of basic equipment, with typical prices.

1   What is the total cost of all the equipment you will need to put together your own biodiesel reactor?

2   How much would you save by building your own processor rather than buying one that is already assembled? Show your answer as a percentage saving.

Task 2

1 a The process for making biodiesel fuel is shown on the data sheet. Ignore the cost of materials for now. Which **three** stages of the process have additional costs?

b The diagram shows the electricity readings at the start and end of the three-hour drying period. The meter measures the units of power, in **kilowatt hours** (kWh). How much power has been used during processing?

c Your processor includes two heaters, each of power 3 kilowatts (kW). They switch on and off to maintain the correct temperature for the process. A 3 kW heater uses 3 kW every hour. For how long were the heaters actually running?

d At stage 8 of the process, the heaters are actually on for 180 minutes during the period from 8:30 pm to 10:15 am the next morning. How much power have you used?

e If electricity costs 14p per kWh, what is the total cost of the electricity used to produce your fuel?

2 The amount of methyl alcohol you will need to add depends how clean your waste oil was.

a If you use 25 litres for every 100 litres of waste oil, show this in a simple ratio.

$$1 \text{ litre} = 1000 \text{ cm}^3$$

b If water costs 113p per cubic metre, what is the cost of water per litre?

c To make 50 litres of fuel you could need 200 litres of water to wash the oil. What is the cost of water needed to wash your oil?

Task 3

Now that you know how to make biofuel, you need to collect **waste vegetable oil** for your processor.

Villages A, B and **F** each have a fish and chip shop that uses vegetable oil and supplies you with 10 litres of waste oil every other week. **Villages C, D** and **E** each have a Chinese takeaway that supplies you with 5 litres of waste oil every week. All distances in the plan below are in miles.

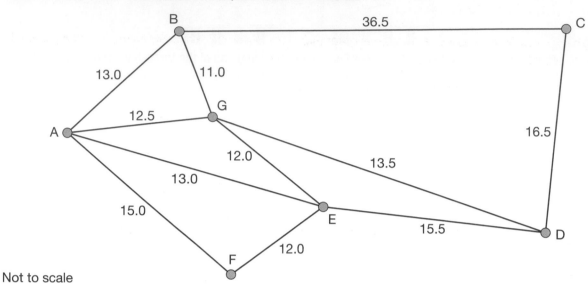

Not to scale

1.
  a. Suppose you live in village G. Plan your shortest route to collect the waste oil.

     How many miles do you travel:

     i. in the weeks you collect from all the shops

     ii. in the weeks you collect only from the Chinese takeaways?

  b. Compare your route with those of others in the class.

2. If your car does 55 miles per gallon (mpg), how much fuel will you use a month to collect your waste oil?

3. If you use a fuel station to fill your car with fuel costing £1.17 per litre, how much will it cost to collect your waste oil?

Task 4

Villages A, B and F each supply you with 10 litres of waste oil twice per month, villages C, D and E each supply you with 5 litres per week. You can process 50 litres of waste oil per batch. Methyl alcohol costs £17.31 for 25 litres.

1. How much waste oil do you collect per month (assume this is a four-week period)?

2. Using your answers from Tasks 2 and 3, calculate the cost to produce a 50-litre batch of biodiesel.

3. Using your answer from question **2**, what does your biodiesel cost per litre?

Task 5

1   a The average person drives 14 000 miles a year. If your car does 55 mpg and diesel costs £1.17 per litre, how much would you spend per year? Check your answer.

     b How much would you spend on fuel if you used your own biodiesel?

     c What are your savings per year?

Soon, more and more people want vegetable oil for making fuel. Your suppliers now charge you 38p per litre for their waste oil.

2   a How much would this increase your annual cost of fuel?

     b How much would you still be saving on fuel, per year?

     c Is it worth producing your own fuel at this rate?

Task 6 (extension)

World consumption of biodiesel – 2007

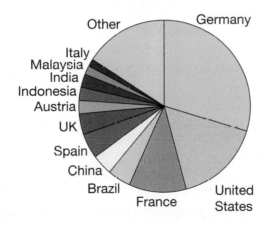

1   The pie chart shows that in 2007, the world's major consumer of biodiesel was Germany. Is this still true today? Investigate and find up-to-date figures. Use an appropriate mathematical chart and write a short report to show your findings.

2   An increasing number of national filling stations now sell a form of biodiesel. How do these prices compare to standard diesel prices? Would there be a significant saving if you used these biodiesels?

HOW DID YOU FIND THESE TASKS?

• What did you find easy or difficult about these tasks?

• Did you work on your own, in pairs or in groups, and how did this help or hinder your approach and success with these tasks?

• What did you learn about how maths is used and applied in real-world situations?

Thumbelina

Thumbelina is a fairy tale written by a Danish author, Hans Christian Andersen, in 1835. It tells the story of an old woman who longs for a child. She receives a magic seed, from which grows a thumb-sized girl, Thumbelina. One night, Thumbelina is snatched by a toad looking for a bride for her son. She manages to escape by floating away on a lily pad and then has a series of adventures involving different animals. Finally, she meets a small flower-fairy prince whom she marries.

This activity is based on a theatre company that has decided to use the story of *Thumbelina* to create a Christmas play. A teenage actress will take the role of Thumbelina. The costumes for the animals she meets, and the set, will be designed so that the world around her is proportionately large.

Task 1

The costume designer has collected the heights of each of the animals cast in the play. She has also measured the height of the actress and the length of her own thumb. All of this information is recorded in data sheet 1, which you have been given.

1 Find the ratio, in the form $1 : n$, of the costume designer's thumb to the height of:

   a the toad   **b** the fish   **c** the butterfly

   d the fieldmouse   **e** the actress.

2 Use the information above and data sheet 1 to work out the height of each of the costumes for the animals.

   Remember: The set is to be designed so that all the animals' costumes are in the same proportion to the actress, as the real animals would be to a thumb.

3 The costume designer is informed that the cast playing the parts of animals includes:

- a woman
- a stilt-walking man
- another man
- a child

Use your answers to question **2** to work out who will play which animal.

Task 2

There are four main scenes in the play. The set designer begins to plan the set for the first three scenes.

Scene 1: Old woman's home

Scene 2: Thumbelina floats away

Scene 3: Mouse hole

Each scene will have its own set. These are all built to the same 3D plan but then decorated appropriately. The set designer's ideas and notes are on data sheet 2.

1 Draw the 3D shapes for each set on an isometric grid, with a scale of 1 centimetre to represent 1 metre.

2 Make a scale drawing of the plan view for each set.

3 Make scale drawings of the front and side elevations for each set.

4 The set designer uses blocks measuring 1 m × 1 m × 1 m that lock together to make each of the 3D shapes.

   a Work out how many blocks are required for each set.

   b Work out how many blocks are required in total.

5 **a** What is the ratio of the number of blocks required for set 1 to the number of blocks required for set 3?

   b What is the ratio of the number of blocks required for set 1 to the number of blocks required for set 2?

6 The producer decides to cut costs and re-use the blocks for set 1 in set 2 and set 3. The re-used blocks will be painted differently on opposite sides and turned around for each set change.

How many blocks are required in total now?

Task 3 (extension)

Scene 4: Flower prince on the hill

Design your own set, made from metre-cube blocks, for this scene. It must be made from the same blocks as are used in sets 1, 2 and 3, so make sure you do not include too many blocks.

Use an isometric grid to draw your set to scale. Then draw the plan, front and side elevations to scale.

Task 4

The lighting designer has lights rigged up around the theatre, at different heights and distances from the stage. He needs to be certain that each one has the correct throw distance to light each scene.

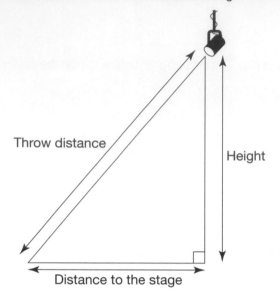

Throw distance

Height

Distance to the stage

1   Find the throw distance for these lights. Give your answers correct to 1 decimal place.

   a   Light A is 16 m from the stage and rigged at a height of 9 m.

   b   Light B is 18 m from the stage and rigged at a height of 10 m.

   c   Light C is 20 m from the stage and rigged at a height of 11 m.

2   Light D is 22 m from the stage and rigged at a height of 12 m. Use your answers to question **1** to estimate its throw distance. Explain how you found your answer.

Task 5 (extension)

Lights E, F and G are on a scaffold placed 10 m from the stage. The throw distance for G is 15.6 m and the ratio of heights E, F and G is 1 : 2 : 3. At what heights are they placed?

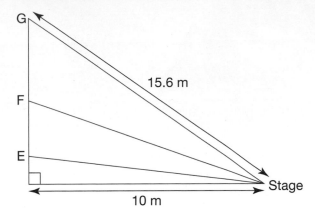

Task 6

Imagine you are the theatre producer for *Thumbelina*. You are responsible for finding the people to work on the show.

Write a job specification for each of the following roles.

1   Costume designer

2   Set designer

3   Lighting designer

Include in each specification:

- a sentence describing the role

- a list of mathematical skills you think are necessary for the role

- any other skills that you think may be essential to the role.

HOW DID YOU FIND THESE TASKS?

- What did you find easy or difficult about these tasks?

- Did you work on your own, in pairs or in groups, and how did this help or hinder your approach and success with these tasks?

- What did you learn about how maths is used and applied in real-world situations?

Off your trolley

Supermarkets have changed the way we shop. Before there were supermarkets, people queued up and bought goods by asking the assistant, who would fetch them from the shelf.

Learning objectives

Representing Level 1: identify and obtain necessary information to tackle the problem

Analysing Level 1: solve problems requiring calculation with common measures, including money

Interpreting Level 1: interpret and communicate solutions to practical problems, drawing simple conclusions and giving explanations

LINKS WITH
Food technology
Science
Economics

Supermarkets were introduced to Britain in about 1960. Shops with wide aisles with everything laid out and pre-packaged allowed everyone to serve themselves.

Supermarkets are owned by large companies, in some cases multinationals, so they can buy in huge quantities and therefore keep prices low.

Task 1

Your **carbon footprint** is the amount of carbon dioxide (CO_2) you generate.

This table lists various food items. The figures show the amount of carbon dioxide that is given off in producing each item, with the average consumption per person per year.

Food	Amount of CO_2 produced per lb of food (in lb)	Average annual consumption per person (lb)
Lamb	16	4
Beef	15	25
Pork	6.75	42
Chicken	3.4	50
Tuna	4.5	0.3
Shrimp	2.7	0.2
Salmon	0.06	0.6
Tomatoes	9	3
Potatoes	0.4	21
Milk (1 pint)	1.5	160

1   How much carbon dioxide is made in the production of one person's annual milk consumption?

2   What is the total amount of carbon dioxide generated in the production of all the foods in the table, just for one person?

3   Apart from those in basic production processes, what other factors add to the carbon footprint of food?

Task 2

Georgie has the choice of two supermarkets at which to shop.

The data sheet lists the prices of the goods Georgie wants to buy.

1   Calculate the cost of the shopping at Supermarket A.

2   Calculate the cost of the shopping at Supermarket B.

3   Calculate the cost if Georgie visits both supermarkets to get the best deal from each.

4   Supermarket A is a three-mile drive each way.

Supermarket B is a seven-mile drive each way.

It is five miles from A to B.

If the cost of driving is 30p per mile, should Georgie go to Supermarket A, B or both?

Task 3

You can save money by buying food on special offer, or by buying larger quantities but, if you buy too much and the food goes mouldy or bad, you are not saving a penny!

Raw chicken might keep in the fridge for two or three days before being cooked, and semi-skimmed milk might last a week. But both can be frozen, as can pizzas and bread. It's a good idea to think about value for money.

Have a look at the prices of these products.

Chicken breasts	333 g for £3	613 g for £6.74	950 g for £10.43
Carrots	500 g for 71p	600 g for 90p	750 g for 92p
Baked beans	220 g for 30p	420 g for 44p	4 × 420 g for £1.48
Orange squash	64 p for 1 litre	£1 for 1.5 litres	£1.35 for 2 litres
Shower gel	150 ml for £1	250 ml for £2.32	400 ml for £2.75
Cheddar cheese	400 g for £3.98	600 g for £4	2 × 400 g for £5

1   For each food item, which size is the best value for money?

2   Which items might be more difficult to keep fresh, if you buy more than you need straight away?

Task 4

Mario is using this recipe. His shopping list shows the prices of the packs sold by his local supermarket.

RECIPE
Chicken and bacon
Serves six
6 rashers bacon
1 tbsp butter (rounded)
1 tbsp olive oil
6 boneless, skinless chicken breasts
1 onion, chopped
3 cloves garlic, minced
$\frac{1}{2}$ cup grated cheddar cheese

SHOPPING LIST

Bacon	10 rashers for £1.64
Butter	£0.98 for 250g
Olive oil	£1.84 for 500ml
Chicken breasts	£3.90 for 4
Onion	£0.50 for 3
Garlic	£0.24 per bulb
Cheddar cheese	£1.50 for 250g

1   Find the cost of all the ingredients. In some cases you will have to buy more than you need.

2   Find the total cost of the meal per person.

3   Use the notes to find a better estimate of the cost of the meal per person.

Notes

1 rounded tbsp butter weighs about 20 g.

1 tbsp olive oil is 15 ml.

3 cloves of garlic are about $\frac{1}{4}$ of a whole garlic bulb.

1 cup of grated cheddar cheese weighs about 100 g.

Task 5 (extension)

1   Find out the cost of a supermarket ready-meal.

2   Find a recipe to make the meal yourself, and find the cost of the ingredients.

3   Is it cheaper to buy it ready-made or make your own?

HOW DID YOU FIND THESE TASKS?

- What did you find easy or difficult about these tasks?
- Did you work on your own, in pairs or in groups, and how did this help or hinder your approach and success with these tasks?
- What did you learn about how maths is used and applied in real-world situations?

The honey bee

The sight of a bee buzzing from flower to flower is typical of long, summer days. The bee, like other insects, plays a key role in the pollination of plants, including flowers and crops, and is thus a key player in our food chain. However, this scene is becoming increasingly rare around the world.

The worldwide decline in the number of **honey bees** is due to factors such as larger fields, leaving fewer wild flowers, and the increased use of pesticides.

Learning objectives

Representing Level 1: find information, from the internet and tables, required to tackle a problem

Analysing Level 1: apply mathematics in an organised way when working with straightforward problems

Interpreting Level 1: interpret results and draw simple conclusions for the problem that is set

LINKS WITH
Science
Geography
ICT
English
PSHE

Task 1

Most honey bees in the UK are **domesticated**, which means that they are looked after by beekeepers.

1 The table shows annual statistics for British beekeepers.

a Using the information in the table, describe the pattern in beekeepers and hives from 2007 to 2010.

	2007–08	2008–09	2009–10
Loss rate (%)	30.1	19.2	17.7
Hives per beekeeper (average)	3.7	3.9	4.7
British beekeeper members	12 500	14 000	17 500

b Predict the number of beekeepers in 2010–11 and in 2011–12.

2 A hive can contain up to 40 000 honey bees.

a Study the table above. How many bees did a typical beekeeper lose for each of the three years?

b Make a table to show the number of bees before and after the loss for each year per beekeeper.

c How many bees were there in domestication by 2010?

3 This table shows regional losses, recorded as percentages across England, for 2009–10.

Region	Loss rate (%)	Region	Loss rate (%)
Northern	26.0	Eastern	17.0
North Eastern	17.9	South Eastern	16.6
Western	18.6	Southern	19.7
South Western	12.8	**National (England)**	17.3

a How does the regional loss rate compare to the National (England) loss rate for 2009–10?

b Using the information in this table, produce a bar chart, add a mean line and comment on your chart.

c Sketch a map of the UK. Divide it into appropriate regions and add the information from this table. Comment on any pattern in loss rates.

Task 2

As well as gathering nectar to produce honey, honey bees perform a vital secondary function, which is **pollination**. About one-third of the human diet comes from insect-pollinated plants, and honey bees are responsible for 80 per cent of this pollination.

1. How much of our diet has been pollinated by the honey bee? Show your answer as both a fraction and percentage.

A honey bee lives from 28 to 35 days. She makes about 24 trips back to the hive every day, visiting from 50 to 100 flowers per trip.

2. What is the maximum and minimum number of flowers a honey bee visits per day?

3. What is the maximum number of flowers she visits in her lifetime?

Task 3

The honey bee quiz

1. A standard honey jar holds 1 lb of honey.

   Honey bees must visit 125 000 flowers to make an ounce of honey.

   How many flowers must honey bees visit to produce 1 lb of honey?

2. A hive of honey bees must fly 88 000 km to collect the nectar to make 1 lb of honey. How many miles is that?

3. What is the shape of a honeycomb?

4. Draw a tessellation of the shape to show the honeycomb.

5. How many teaspoons of honey do you put on your toast?

   The average worker bee makes $\frac{1}{12}$ teaspoon of honey in her lifetime. How many worker bees does it take to make the honey for your toast?

6. A honey bee weighs 0.218 g. A bee's load of pollen weighs 10 mg.

   Write the ratio of the weight of the honey bee to the weight it carries. Comment on your answer.

7. What would be the equivalent weight you would have to carry if you were a honey bee?

Task 4

Imagine you were a honey bee. Using the honey bee map on data sheet 1, plan a route to collect as much nectar as you can each day. You may not visit more than 100 flowers per trip and are allowed no more than 24 trips – to the flowers and back to the hive – in a day. Each flower may be visited only once. Plan a route that involves the minimum flying distance.

1   Based on your calculations from the map, how many days will it take you to collect all the nectar? Keep a record of the distance travelled in each trip.

2   How far will you have travelled to collect all the nectar?

3   Compare your answer to those of other members of the class. Did you have the shortest route?

Task 5 (extension)

The honey bee is a mere half-inch long, from front to back. Its wings flap about 180 times per second.

1   a   How many times per minute do its wings flap? Give your answer to the nearest thousand.

    b   If a honey bee flies around collecting nectar for six hours per day, how many times do its wings have to flap while it is collecting nectar? Give your answer to 2 significant figures.

2   a   Using this image, calculate the length of the honey bee. Give your answer in millimetres.

    b   Produce a scale drawing of the honey bee. Use a whole sheet of A4 paper. Choose a suitable scale.

Shown twice actual size

    c   Use your scale drawing to find the length of the wings of the honey bee. Give your answer for a real bee.

HOW DID YOU FIND THESE TASKS?

- What did you find easy or difficult about these tasks?
- Did you work on your own, in pairs or in groups, and how did this help or hinder your approach and success with these tasks?
- What did you learn about how maths is used and applied in real-world situations?

Money – making the world go round?

The coins used in the UK are made by the Royal Mint, which has its headquarters in South Wales. It is owned by the British Government. Each year the Royal Mint makes coins for about 60 countries, including the UK.

There are eight different denominations of coin in circulation in the UK. The total number of coins is about 28 billion (28 000 000 000). Great care is taken to make sure every coin is perfect.

In the past, coins were made of gold and silver but now even our 'copper' coins (1p and 2p) are no longer real copper, because of the high cost of the metal. They are made of steel, plated with copper. The steel makes them very strong. The 'silver' coins (5p, 10p, 20p, 50p) are made of an **alloy** of copper and nickel. The Royal Mint does not reveal the exact percentages of each metal in the alloy. The 'gold' coins (£1, £2) are made from an alloy of three metals.

Some coins in circulation are illegal or **counterfeit**. Figures suggest that up to 2.5% of our coins could be counterfeit.

Learning objectives

Representing Level 1: obtain the necessary information to tackle a problem; draw bar charts

Analysing Level 1: solve problems requiring calculation, with common measures, including money, length and weight; work out areas of rectangles and calculate percentages

Interpreting Level 1: extract and interpret information from tables; compare bar charts

LINKS WITH
Citizenship
Science
Economics
English

Task 1

The table below shows the numbers of each denomination of coin in circulation.

Denomination	Number of coins (millions)	Value (£millions)
£2	345	690
£1	1474	
50 pence	845	
20 pence	2473	
10 pence	1651	
5 pence	3774	
2 pence	6664	133.28
1 penny	11 215	
Total	**28 441**	

1   Complete the table to find the total value of all the coins in circulation. Then round your numbers to the nearest million.

2   Draw a bar chart to show the **numbers** of each type of coin in circulation.

3   Draw a bar chart to show the **total value** of each type of coin in circulation.

4   Compare the shapes of your bar charts.

Task 2

The £1 coin was introduced on 21 April 1983 to replace the £1 note.

The £1 coin is minted from an alloy that is about 70% copper, 24.5% zinc and 5.5% nickel.

Each £1 coin weighs 9.5 grams and has a diameter of 2.25 cm.

1   Calculate the total weight of all the £1 coins in circulation.

2   Calculate how much of the total weight you have found in question **1** is copper, how much is zinc and how much is nickel.

Task 3

It has been estimated that the cost of the metal in a £1 coin is about 4p.

Counterfeiters make illegal coins. It is difficult to be sure how many counterfeit £1 coins are in circulation but it is estimated that 2.5% of all £1 coins are fake. In July 2010 the Royal Mint considered discontinuing the £1 coin because of the increasing number of counterfeits.

After the metal is prepared in flat sheets, it is cut into discs, called **planchets**, which are exactly the right size for the coins being made. The next step is to **strike** the coins, to put on the detail.

The £1 coin uses exactly the same planchet as the Swazi **lilangeni** (the main unit of currency in Swaziland). Both coins (the £1 and the lilangeni) are the same size and made of the same alloy, but the lilangeni is only worth about 9p. It is relatively easy to pass off a lilangeni as £1, as it will be accepted by all UK vending machines as both coins are the same size and weight.

1 How many £1 coins are fake, according to the above-mentioned estimate?

2 What is the value of 200 £1 coins?

3 What is the value of 200 lilangeni?

4 Draw a conversion graph to show the conversion of pounds to lilangeni.
Your graph should go up to 200 lilangeni.

Task 4

The planchets are cut from a strip of metal exactly the right width to avoid waste. To make £1 coins of diameter 2.25 cm, a metal strip 2.25 cm wide is used.

Occasionally, the Royal Mint makes commemorative £5 coins, which have a diameter of 3.86 cm. The face of a £1 coin has an area of 4.0 cm^2 and a £5 coin has a face area of 11.7 cm^2.

1 How long would the strip need to be to make 1000 £1 coins?

2 What is the area of the top surface of the same strip?

3 What is the area of metal actually used to make 1000 £1 coins?
Use the area of one £1 coin, given above.

4 What percentage of the metal strip is used for the coins?

5 How wide would the strip need to be to make £5 coins?

6 How long would the strip need to be to make 1000 £5 coins?

7 What is the area of the top surface of this strip?

8 What is the area of metal actually used to make 1000 £5 coins? Use the area of one £5 coin, given above.

9 What percentage of the metal strip would be used to make 1000 £5 coins?

Task 5 (extension)

There are several methods of spotting fake £1 coins.

- The image might not be as sharp as on a real coin.

- The metal alloy might be different, so the colour or weight might not be exactly right.

- The reverse or 'tails' side of the £1 coin changes every year. On many counterfeit coins, the **obverse** or 'heads' side, which includes the date, and the image for the year do not match.

- The obverse and reverse sides should be aligned, so that the top of each side matches. Many counterfeits do not match in this way.

Counterfeit coins are worthless. However, genuine mistakes, made by the Royal Mint, can be worth more than the face value. For example, in 2008 the Royal Mint produced a small number (less than 250 000) of 20p coins without including the year in the design. These currently sell for over £60 because of their relative rarity.

1 Research the different designs that have been on the reverse of the £1 coin since 1983.

2 Research other coins that are worth more than their face value.

HOW DID YOU FIND THESE TASKS?

- What did you find easy or difficult about these tasks?
- Did you work on your own, in pairs or in groups, and how did this help or hinder your approach and success with these tasks?
- What did you learn about how maths is used and applied in real-world situations?

Flags

Flags have been used for thousands of years. The first known use was by armies, as standards. Then they were used by sailors, at sea, and now they have many different purposes.

Flags are usually designed so that they can be easily seen from a distance. They need to be recognised and understood even when they are being waved or moved, for example, in the wind. The designs of most flags include colour and many include symmetry.

Task 1

The table opposite shows some flags that are used as warnings on beaches in different European countries.

1   For each of the flags in the table, find:

   a the number of lines of symmetry

   b the order of rotational symmetry.

2   **a** Look at Flags 1–5, which are all shown the correct way up. Would you be able to tell if the lifeguard were to hang them upside down? Which ones would look different?

   b What do you notice about the order of rotational symmetry of those flags that would look the same if they were hung upside down?

3   This is a jellyfish warning flag, used on Spanish beaches.

   Write three sentences commenting on its symmetry and design. Explain whether you think its design is good or bad, giving reasons for your answer.

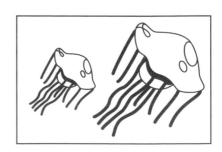

Flag		Country and meaning
Flag 1		**UK:** Lifeguards patrolling Safe to swim
Flag 2		**UK:** Area used by watercraft (such as windsurfers and kayaks) Not safe to swim
Flag 3		**UK:** Watercraft use prohibited
Flag 4		**Portugal:** Beach temporarily without a lifeguard
Flag 5		**France:** Windsurfing dangerous due to strong winds or sea conditions
Flag 6		**France:** Swimming and use of floating devices hazardous because of land breeze
Flag 7		**France:** Surfing area

Task 2

Semaphore is a system of signalling, using the positions of flags. It is useful when electronic communications may be difficult, for example, in the mountains.

Each letter of the alphabet is represented by two flags, one held in the signaller's right hand and one held in the left hand. For example, this is the signal for the letter P.

Imagine that the signaller can only use the positions described. In each case, each letter is represented by two flags, one in the right hand and one in the left hand.

1 Two right and two left

   a How many possible signals can the signaller make?

   b How many more letters than this are there in the alphabet?

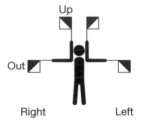

2 Three right and two left

   a How many possible signals can the signaller make now?

   b How many more letters than this are there in the alphabet?

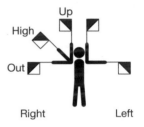

3 Four right and two left

   a How many possible signals can the signaller make now?

   b How many more letters than this are there in the alphabet?

   c What do you notice about the number of possible signals, compared to the numbers of right positions and left positions?

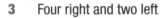
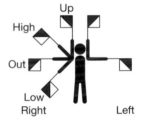

These are all of the possible semaphore positions.

Note: The down signal can only be made by one hand at a time (right or left). This is different from the up signal, which can be made by both hands (right and left).

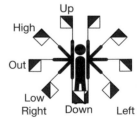

4 How many different right positions, including down, are there?

5 How many different left positions, not including down, are there?

6 How many possible signals can the signaller make altogether?

7 What might you do to create the signals for the remaining letters in the alphabet?

8 Look at data sheet 1. What does the signaller do to create signals for the letters that have not been included in the signals you have seen? **Hint:** Look at letters H, I, O, W, X, Z.

9 Write two sentences to say whether you think the semaphore system is a good or a bad way of sending messages. Give reasons for your answer.

Task 3 (extension)

Use semaphore to sketch a short message to a friend.

Task 4

Look at data sheet 2, which gives the overall proportions of the national flags of the countries in the UK.
You should see that the normal proportions for height to length are 3 : 5.

1   Ella has a sheet of A4 paper. She wants to draw the flag for England to scale, but it won't fit.

   a   She thinks if she divides all the 'real' dimensions by the same number, she could make it fit, without making it too small. What number should she divide by?

   b   Sketch the flag with the dimensions Ella will use on it.

2   Lily has a sheet of A5 paper. She wants to draw the flag for Scotland to scale, but it won't fit.

   a   What number could she divide the dimensions on the diagram by, to make it fit, without making it too small?

   b   Sketch the flag with the dimensions Lily will use on it.

3   Mandy has a sheet of A3 paper. She wants to draw the Union flag with the dimensions exactly as on data sheet 2, in centimetres, but she needs some help.

   a   She thinks it will be too difficult to draw an angle of 30.96°. What angle do you think Mandy should draw to make it easier?

   b   Mandy doesn't know how far along to draw the three stripes labelled 2, 6 and 2, below. They need to be in the middle of the flag.

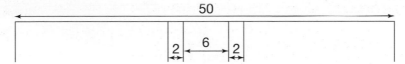

Sketch this part of the flag with the dimensions on it to show Mandy how far along the stripes should be.

   c   Mandy has the same trouble with these stripes labelled 2, 6 and 2, on the right. Again, she needs them to be in the middle of the flag.

   Sketch this part of the flag with the dimensions on it to show Mandy how far along the stripes should be.

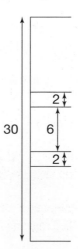

4   Mark measures the picture of the Welsh flag on data sheet 2. He decides to enlarge it to A2 on a copy shop's photocopier. He wants to make it as large as possible, but doesn't want to lose any of the picture.

Which option should Mark choose on the photocopier?

×2   ×3   ×4   ×5   ×6   ×7

Explain your answer, giving the dimensions of Mark's flag.

5   Look very carefully at each flag. Write two sentences about each one, commenting on its lines of symmetry and order of rotational symmetry. Say whether you think it is a good or bad design for a flag, giving reasons for your answer.

Task 5

Flagpoles are usually too high to measure simply and accurately, but this task shows you a way to estimate the height. You will need two metre rulers, a set square or a protractor and a sunny day!

- Find the shadow of the flagpole. Now ask a partner to hold a metre ruler so that it touches the ground, somewhere close to the flagpole. Use a set square or a protractor to make sure it is perpendicular to the ground. The shadow of the metre ruler and the shadow of the flagpole should be parallel.

- Use the other metre ruler to measure the lengths of the shadows of the ruler and the flagpole.

- The proportion of the heights to the shadows will be exactly the same:

$$\frac{\text{height of flagpole}}{\text{length of flagpole's shadow}} = \frac{\text{height of ruler}}{\text{length of ruler's shadow}}$$

- Then you can work out the height of the flagpole:

$$\text{height of flagpole} = \text{length of flagpole's shadow} \times \frac{\text{height of ruler}}{\text{length of ruler's shadow}}$$

1   Councillor Murray is doing a survey of the heights of flagpoles in his town, so he knows what lengths of rope will be required when the hoisting equipment needs replacing. He uses the method shown above and sketches his findings.

Use the sketches to calculate the height of each flagpole.

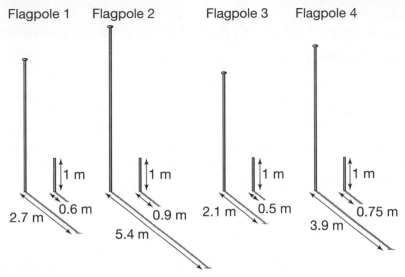

Flagpole 1   Flagpole 2   Flagpole 3   Flagpole 4

1 m   1 m   1 m   1 m

2.7 m   0.6 m   0.9 m   2.1 m   0.5 m   0.75 m

5.4 m   3.9 m

2   Councillor Murray decides to write a report, including the heights flags should be raised on the flagpoles in different circumstances. For example, when a very important person dies, such as a member of the Royal Family, the flags should be lowered to half-mast. This means the top of the flag is two-thirds up the flagpole.

How far up the flagpole is half-mast for each of the flags in question **1**?

Task 6 (extension)

Ella, Lily and Mandy decide to make flagpoles out of bamboo canes and fly their flags (see Task 5) in the garden. They each have a choice of a 60 cm, 90 cm or 150 cm cane. Choose a cane for each girl's flag. Sketch each of the flags flying:

1   at the top of its flagpole

2   at half-mast.

Include dimensions on your sketches.

Task 7

A children's charity is running a competition for a well-designed flag to represent it. The flag will be flown on the rooftop of the charity's headquarters. It would also like the position of the flag on the flagpole to communicate a message.

Design and draw a flag, including the dimensions. Then decide what message the flag could be used to send by its position on the flagpole. Write a paragraph to explain.

HOW DID YOU FIND THESE TASKS?

- What did you find easy or difficult about these tasks?
- Did you work on your own, in pairs or in groups, and how did this help or hinder your approach and success with these tasks?
- What did you learn about how maths is used and applied in real-world situations?

Commuting

Official statistics for 2009 reveal that workers in the UK spend nearly 22 million hours commuting every day, which gives a daily average of 52.6 minutes. Workers in well-paid jobs, such as managers and senior officials, have the longest commute times of 68.6 minutes; those in lower-paid occupations, such as cleaners and labourers, commute for 40.4 minutes on average.

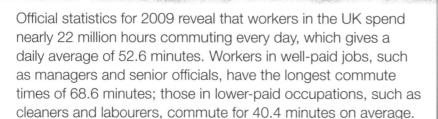

★

Learning objectives

Representing Level 1: obtain information from tables and the internet to tackle the problem

Analysing Level 1: apply mathematics in an organised way

Interpreting Level 1: interpret results in the form of graphs and charts, giving a simple explanation

LINKS WITH
Geography
ICT

Task 1

In autumn 2007, there was a survey on commuting patterns in the UK. This identified nine categories for modes of transport, divided into different areas of London, as well as Britain as a whole. The map shows London divided into the regions used in the survey.

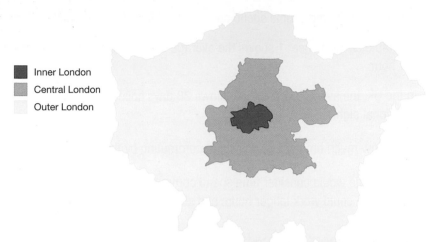

■ Inner London
■ Central London
□ Outer London

The table shows the percentages of people using different modes of transport. Use the map to identify the three different areas of London.

Main mode	Area of workplace					
	Central London	Rest of inner London	Outer London	All London	Rest of Great Britain	Great Britain
Car and van	11	31	63	37	76	71
Motorbike, moped, scooter	2	1	1	1	1	1
Bicycle	3	4	2	3	2	3
Bus and coach	12	16	14	14	7	7
National Rail	40	16	4	19	2	4
Underground	27	19	5	16	–	2
Walk	4	12	10	9	11	11
Other modes	1	1	1	1	1	1
Number of people (millions)	1.11	0.87	1.36	3.34	21.48	24.83

1 a What percentage of people in central London used public transport to get to work?

   b How many people in central London used public transport to get to work?

2 Information is usually easy to understand when it is shown in a chart or graph.

   a Draw a pie chart to show the proportion of people using each mode of transport to get to work for 'All London'.

   b Calculate the number of people that commuted to work for each mode of transport and add this information to your pie chart.

   c Draw a pie chart to show the same information for 'Rest of Great Britain'. Comment on any similarities and differences.

★ Beginner

Task 2

The amount of time people spend commuting affects their work–life balance.

The table on data sheet 1 shows the average travel time to work and back, by occupation, and average pay per hour.

In the UK, the average occupation allows 28 days holiday per year. All public and bank holidays account for an additional eight days.

1 How much time per year is spent commuting by the average worker in each occupation?

Most people would consider time spend commuting 'work' time rather than 'leisure' time. Someone working from home could work longer hours and still have the same work–life balance as someone who commutes to work.

2 **a** People are not usually paid for the time they spend commuting. Use your answer to question **1**, and the average pay per hour given in the table, to work out what the average yearly cost of commuting for each occupation would be, if employees were paid for this time.

    b If the total number of hours spent commuting per day is 21.8 million, and the average hourly pay is £13.90, what would be the total cost of commuting per day?

Task 3

Many people in the UK now work from home and commute into work only two or three days a week. Even people living in the north of England commute to London a couple of days a week.

1 **a** Using the mileage chart on data sheet 2, find the distance between London and each of the following places.

    i Luton   **ii** Birmingham   **iii** Manchester

    iv Norwich   **v** Newcastle

  b Assuming an average speed of 50 mph, how long would it take to drive to London to each of the places in part **a**?

  c The cost of fuel is £1.20 per litre. Assume your car does 45 miles to a gallon.

> 1 gallon is equivalent to 4.55 litres.

  Calculate the cost of fuel for each journey in part **a**.

2 Using the table on data sheet 2, compare the time and cost of commuting to London by car and by train from each of the five places in question **1**.

Task 4 (extension)

In 2003, when house prices in the south of England rose beyond the budgets of many people, more than 10 000 people were offered an alternative when Kent County Council started to draw up plans with Eurotunnel to encourage people to buy property in the Calais area of France.

1 **a** Find the travel time and cost of commuting from France to London.

  b Is the idea of living in France and working in the UK reasonable?

2 Budget airlines now offer very cheap flights. How does commuting from France to London compare to what you found out in Task 3? Where would you rather live?

HOW DID YOU FIND THESE TASKS?

- What did you find easy or difficult about these tasks?
- Did you work on your own, in pairs or in groups, and how did this help or hinder your approach and success with these tasks?
- What did you learn about how maths is used and applied in real-world situations?

Website design

http://www.business-...

Whenever we use the internet, we are actually using a website. A search engine, a company's home page and a social networking site are all just websites. Even when you check sports results, you are using a website.

Nowadays, most businesses have their own websites, but what does it cost to set one up and run it? Suppose a plumber decides to have a website, called www.round-the-bend.co.uk. In this activity, you will find out how he sets about it.

Task 1

Before you start to design a website you need to plan it and decide how many pages or screens it will have. You could choose to have a home page, a page describing your company and another about what you are offering. The number of pages will determine how much someone will charge you to design your website.

Imagine that you are starting a small business that sells cupcakes. You decide that you need a website so that people will know what you are offering and how to find you.

1   a   Make a list of pages you will need. How many pages will your website have? Choose appropriate headings, although you don't need to add any details yet.

    b   Draw a flow chart to show all the pages and how they will link to each other.

2   You will need to choose a host and package for your website.

    a   Use data sheet 2 to choose a suitable company to host your website, and the best package of services. Write down the details and calculate how much it will cost per year.

    b   You see two companies advertising web design. Work out how much Media Design would charge to produce your website.

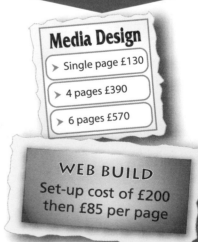

Media Design
> Single page £130
> 4 pages £390
> 6 pages £570

WEB BUILD
Set-up cost of £200 then £85 per page

Task 2

Now you are ready to design the website. Before you start, think about some websites you have used. Some websites are badly designed, for example, when they don't fit on the screen properly. Most computer monitors are classified as 15 inches. This is a measure of the distance diagonally across the screen.

1 inch ≈ 2.54 cm

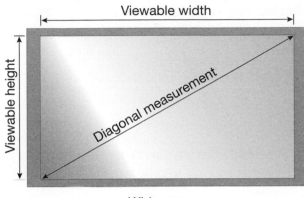

Widescreen

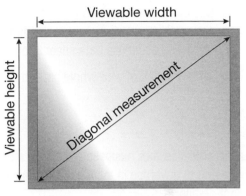

Standard

1   a   Mark a diagonal line across a sheet of A3 paper, in landscape orientation. Convert 15 inches to centimetres and mark this length along the diagonal, from one corner. Join the line to the horizontal and vertical sides of the paper as shown in this diagram.

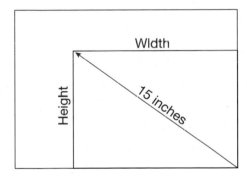

    b   There are 59 pixels per centimetre. What will the dimensions of your webpage be, in pixels?

    c   Remember that the aspect ratio is the ratio of the width of the screen to the height. Will your aspect ratio be similar to a standard 4 : 3 or widescreen 16 : 9?

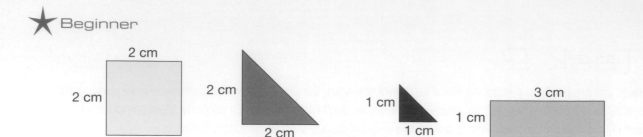

2 Some websites have a company design or logo on all the pages. Try designing your own logo.

a Use centimetre-squared paper to design a pattern that covers an area of 28 cm by 3 cm, at the top of your webpage. You can only use squares, rectangles and triangles, tessellating them so there are no gaps. You could use data sheet 3 for this.

b What fraction of your whole pattern does each type of shape represent?

Task 3

You can publicise your business website in many ways. Two very effective ways are to have it displayed on your vehicle and to advertise in the local newspaper. You will want to include your website address and perhaps other contact details.

1 Based on the two options shown here, how much will advertising cost you for a month?

2 Which of the methods above would you choose to publicise your website? Give a reason for your choice.

Task 4

You now have all the information you need to find the cost of designing, producing and publicising a website.

1 Using the information you have found in Tasks 1 to 3, work out the cost of setting up your website from start to finish.

2 Write a commentary, prepare a slide presentation or produce a short documentary to explain the decisions you have made to evaluate your findings.

HOW DID YOU FIND THESE TASKS?

- What did you find easy or difficult about these tasks?
- Did you work on your own, in pairs or in groups, and how did this help or hinder your approach and success with these tasks?
- What did you learn about how maths is used and applied in real-world situations?

Eat well, live longer?

To keep our bodies healthy, we need to follow a well-balanced diet. We should include food from all the types – **proteins**, **carbohydrates** and **fats** – but in the correct proportions. Proteins are essential to our bodies for maintaining and producing cells and building muscles. They are found in meat, fish, poultry, eggs, milk and milk products. Vegetarians get their protein from sources such as pulses, seeds, nuts, soya and some dairy products. Carbohydrates provide energy. Fats are important because they contain vital vitamins. The difficult part is getting the balance right, to maintain good health and a suitable body weight. Too much, and we become overweight, too little, and we risk becoming undernourished.

In the UK, there is an abundant supply of the 'right' sorts of food. People who live in some other countries are not so lucky.

Learning objectives

Representing Level 1/2: find required information from appropriate tables and choose from a range of mathematics to find solutions

Analysing Level 1/2: work through calculations for mean and range, calculate percentages and use appropriate checking procedures to see that answers make sense

Interpreting Level 1/2: use your results to produce various bar charts, explain what you have found and draw conclusions

LINKS WITH
Science
Geography
English
Media studies
ICT

Task 1

In 2009, the **population growth rate** throughout the world was recorded as 1.133%. Population growth is affected by births and deaths. The rate at which people die is affected not just by natural ageing but also by factors such as medical care and general hygiene. This table shows birth and death rates for 1975 and 2007.

Useful formula: increase or decrease percentage $= \dfrac{\text{change in value}}{\text{original value}} \times 100\%$

| | (Per 1000 of population) | | | |
| | Birth rate | | Death rate | |
Country	1975	2007	1975	2007
Australia	16.9	12.0	7.9	7.6
Belgium	12.2	10.3	12.2	10.3
Czech Republic	19.6	9.0	11.5	10.6
Greece	15.7	9.6	8.9	10.3
Israel	28.2	17.7	7.1	6.2
UK	12.5	10.7	11.9	10.1
USA	14.0	14.2	8.9	8.3

1 Use the information in the table above to copy and complete the table below. Give your answers for the difference in birth rate and death rate correct to one decimal place. Show all other values as whole numbers.

| Country | Difference in | | Population difference per 1000 | |
	birth rate (%)	death rate (%)	1975	2007
Australia				
Belgium				
Czech Republic				
Greece				
Israel				
UK				
USA				

2 Explain your results and give a reason for the negative values in your table.

3 What are the implications of the UK figures?

Task 2

Birth rates and death rates only show a small part of the bigger picture of how people live and how diet affects their **life expectancy**. Life expectancy is the number of years a person is expected to live, basd on statistical probabilities.

1   What do you think the average life expectancy is for people living in the UK, the USA and a country in the developing world?

2   Now use data sheet 1 to check your answers. How close were your guesses?

3   Produce a comparative bar chart to show the life expectancy of both males and females for the following countries: Afghanistan, Andorra, Australia, Brazil, Central African Republic, Kenya, UK, USA and Zimbabwe.

4   What does your bar chart show? Comment on any trends or patterns.

Task 3

The food guide pyramid represents how much of each type of food we should consume per day to follow a well-balanced diet.

1   Using the minimum servings listed in the food guide pyramid, find the minimum number of food servings for a day. Show each food type as a fraction of the minimum number of servings.

2   a A typical recommended serving is 80 g. What is the minimum combined weight of each food type you should eat in a typical day?

   b What is the maximum combined weight of each food type you should eat in a typical day?

3   Draw a food pyramid showing the types of food you eat in a typical day.

4   Using your food pyramid, and referring to data sheet 2, calculate each type of food you eat as a fraction of what you actually eat in a day. Try to do this by weight.

5   In groups of four, discuss ways we can stay healthy. Produce a cartoon-style storyboard or slide presentation and report back to the class.

Fats, oils and sweets
Use sparingly

Dairy
2–3 servings

Proteins
2–3 servings

Vegetables
3–5 servings

Fruits
2–4 servings

RICE

Bread/grains
6–11 servings

HOW DID YOU FIND THESE TASKS?

- What did you find easy or difficult about these tasks?
- Did you work on your own, in pairs or in groups, and how did this help or hinder your approach and success with these tasks?
- What did you learn about how maths is used and applied in real-world situations?

Coastguard search

The sea attracts thousands of people for holidays and weekend breaks as well as sporting events.

Although sea travel is one of the safest forms of transport, there are occasional accidents, mechanical breakdowns, medical emergencies and bad weather that can lead to **incidents**.

Task 1

This table shows the numbers of people assisted, in some way, by the **search and rescue** services around our coasts over a three-year period. It also shows the types of incident.

Type of incident	Year 1	Year 2	Year 3
Leisure diving	39	38	40
Diving – medical	54	101	102
Water sports	270	612	753
Inshore vessel	561	823	738
Commercial vessel	830	550	582
Fishing vessels	521	628	360
Commercial diving	131	121	90
Shore-side	744	512	591
Beach incidents	672	604	705
Other medical assistance	481	431	458
Total	**4303**	**4420**	**4419**

The total number of incidents for each year increased and then levelled off. However, the frequency of some types of incident is decreasing; of others it is increasing.

1   For each type of incident, calculate the percentage frequency change (increase or decrease) from year one to year three.

2   Why did the number of people involved in incidents related to fishing vessel decrease rapidly from year two to year three?

3   Why has the number of people involved in water sports incidents increased rapidly?

Task 2

The data sheet gives details of participation in water-related activities in 2007.

1   Read each statement below, then say whether it is true or false. You will need to refer to the data sheet.

   a More people participate in power-boating than in rowing.

   b Coastal walking is the only activity with more female than male participants.

   c 87% of shore anglers are male.

   d More than half of canoeists are aged between 16 and 34.

   e More than half of windsurfers are aged between 35 and 54.

2   a What are the **two most popular** activities?

   b For each of these two activities, what percentage participate in the UK?

3   a What are the **two least popular** activities?

   b Why do you think they are less popular than other activities?

4   This bar chart shows the breakdown of participants in coastal walking.

   a Produce an appropriate chart to show participation in waterskiing.

   b Describe the main differences between levels of participation in each activity.

5   The notes at the bottom of the table on the data sheet show that some figures have been included more than once. Find the total number of fatalities in the SAR area in 2007, and calculate the number of deaths per million participants.

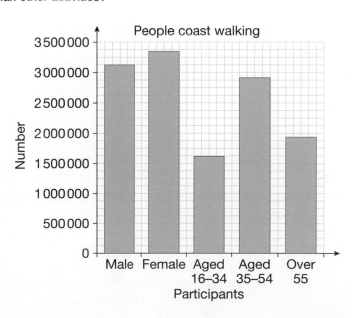

Task 3

Helicopters are often used to search for people who are in trouble in the sea. The helicopter flies high enough to view as large an area as possible, but can also fly low enough to see a person in the water.

There is an **optimum height** at which a helicopter can search most effectively. This depends on visibility and sea conditions, as well as the surveillance equipment being used and the experience of the crew.

The area visible from the helicopter is called its **search footprint**. This is roughly circular but, for this task, you can use a square to make calculations a bit easier.

This square represents a square 1 kilometre by 1 kilometre (area 1 km^2). Each small square represents an area 100 metres by 100 metres.

A helicopter starts from where the incident happened. For example, a search for a man overboard would start from the boat and spiral outwards, covering the maximum area in the smallest possible time.

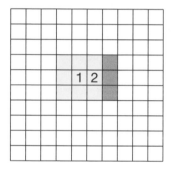

The square marked 1 is the position of the helicopter. The yellow squares show the area where the helicopter can see anyone in the sea. The square marked 2 shows the new position of the helicopter and the green squares show the extra search area.

1   Draw the helicopter's optimum spiral search pattern on a square grid.

2   Copy and complete this table to show the area covered at each stage of the helicopter's search.

Helicopter's position (square number, h)	Number of squares searched (n)
1	9
2	
3	

3   Add a column to your table and work out the total area, in square kilometres (km^2), that will have been covered, from each position.

4   Find a formula or rule connecting h and n.

Task 4 (extension)

A better model for search patterns is to use a hexagonal grid, as shown. The search footprint is shown in colour, with the helicopter in the central hexagon. It is closer to a circular shape (the real footprint).

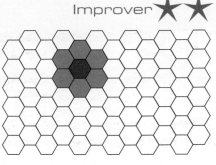

1   Use this new model to create a spiral search pattern. Produce a table, as in Task 3, and find a rule connecting h, the hexagon number, and n, the number of hexagons covered.

2   In the grid, the length of the long diagonal of each hexagon is 100 metres. Calculate the area of the hexagon.

3   What is the total area of the helicopter's search footprint?

Task 5 (extension)

Sea King helicopters are equipped with **synthetic aperture radar**. This is a type of radar that can be used by a moving object, such as a helicopter or an aeroplane.

The radar beam is transmitted at an angle of 45° and has a range of 30°. This is shown in yellow in the diagram.

The beam spreads out to cover a trapezium, shown in green.

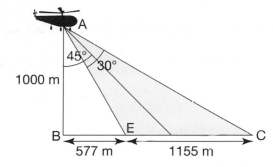

1   Calculate the area of the trapezium.

The helicopter, at A, sees an object at D.

2   Calculate the distance AD.

If the helicopter ascends to a height of 2000 m, all the lengths in the diagram are doubled.

3   Calculate the area of the trapezium in this case.

4   If the area of the trapezium increases as the helicopter flies higher, why should the helicopter not fly as high as possible so that it can scan the maximum area?

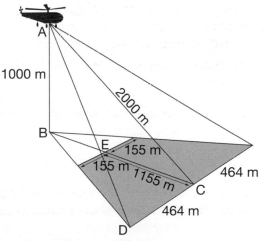

HOW DID YOU FIND THESE TASKS?

- What did you find easy or difficult about these tasks?
- Did you work on your own, in pairs or in groups, and how did this help or hinder your approach and success with these tasks?
- What did you learn about how maths is used and applied in real-world situations?

Photography: past, present and future

The word **photography** comes from the Greek words for 'light' and 'draw'. The first photograph is said to have been taken in the late 1820s by a French inventor, Nicéphore Niépce.

Early cameras were wooden boxes, with single holes or **apertures** that let light in. A plate of special material at the back of the box reacted when it was exposed to light. The image, once developed, was known as a **negative**. The light and dark areas of the picture were reversed. The negative could then be used to print a photograph. This method was refined over the years. The plate was replaced by a roll of film. The basic technique is still used in some cameras today. The first photograph took eight hours to expose.

In modern digital cameras, the film at the back at the camera has been replaced with a light sensor that stores the picture information digitally.

Task 1

Photographers used to **develop** their own film, using chemical solutions in a **darkroom**. Many professional photographers still do this. They say that developing is as much of an art as actually taking the photograph.

Developing a black-and-white film requires several processes before the cleaning stage. Each stage requires a different solution.

> One gallon is equivalent to 4.55 litres.
> 1 litre = 1000 ml

Learning objectives

Representing Level 1/2: identify the problem and find information needed to solve it, using appropriate mathematical methods

Analysing Level 1/2: apply a range of mathematics to find solutions and check that answers make sense

Interpreting Level 1/2: use results to produce graphs and charts, explain what has been found and draw conclusions to justify answers

LINKS WITH
ICT
English
Media studies
Science
History

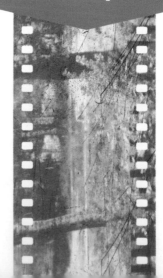

1 a From the list of solutions, how many millilitres of each chemical are needed to make a gallon of solution for the developer (stock solution), stopbath, fixer and rinsing aid?

   b Just before use, the developer (stock solution) is diluted with water in the ratio of 1 : 7. How many millilitres of developer are there in one gallon of the diluted developer solution?

Solutions

The ratios show the proportions of chemical to water.

Developer 1 : 3 (stock solution)

Stopbath 1 : 63

Fixer 1 : 4

Rinsing aid 1 : 400

Hypo-cleaning agent 500 g is mixed with water to produce 19 litres

Store solutions in gallon containers.

2 a A photographer needs one gallon of each solution, plus 19 litres of hypo-cleaning solution, to develop some photos. Using the information in the shopping list, how much will this cost the photographer?

   b The photographer develops two films (24 photos per film) using these chemicals. How much does each photograph cost to develop?

SHOPPING LIST

Developer	(1 litre)	£7.49
	(5 litres)	£24.98
Fixer	(1 litre)	£3.49
	(5 litres)	£14.99
Stopbath	(1 litre)	£3.49
Rinsing aid	(500 ml)	£4.58
Hypo-cleaning agent (500 g)		£5.42

Task 2

Using modern digital cameras, we can print out just the photos we want, at home on a colour printer or dedicated photo printer. For a really professional finish, high-street stores offer cost-effective photo printing, straight from a camera's memory card.

Cost of printing photographs on the high street

		1 hour	24 hour
	0–49	25p	20p
Number of	50–99	22p	15p
photographs	100–149	16p	10p
(6 × 4 inch)	150–199	16p	7p
	200+	16p	5p

Dedicated photo printer costs

Photo printer £149.99

150 sheets of 6 × 4 inch photo paper £39.85

Printer cartridge (black) £19.99
– prints, on average, 600 photos

Printer cartridge (tri-colour) £26.99
– prints, on average, 300 photos

1 a Draw a graph to investigate the cost of printing photographs using a dedicated photo printer.

   b Compare the cost of printing digital photos at home with the cost of having them printed on the high street and the cost of developing film at home, which you found in Task 1.

Task 3

Good photographers and artists position the focus of their pictures in imaginary rectangles with sides in the ratio of 1 : 1.618, both horizontally and vertically. This is known as the **golden ratio**. It is a ratio that is found in many works of art, and in architecture, because the proportions are pleasing to the human eye.

The ratio of 1 : 1.618 sounds complicated but you can use the lazy **rule of thirds**.

1   a   Divide the photographs opposite into three equal parts, in both the horizontal and vertical axes. Do the photographs follow the rule of thirds?

    b   Find other photographs on the internet or at home. What do you notice? Do photographs that use the rule of thirds, or the golden ratio, look better than those that don't?

HOW DID YOU FIND THESE TASKS?

- What did you find easy or difficult about these tasks?
- Did you work on your own, in pairs or in groups, and how did this help or hinder your approach and success with these tasks?
- What did you learn about how maths is used and applied in real-world situations?

Teabag design and production

According to the UK Tea Council, 165 000 000 cups of tea are drunk each day in Great Britain. As well as being a refreshing drink, tea is believed to have many health benefits.

Teabags were first introduced to Great Britain in 1953. Originally, they were rectangular or square. Then, in 1992, circular teabags were introduced to match the shape of a cup or mug. More recently, pyramidal and tetrahedral shapes of teabag have become popular.

Here are some teabags, with their dimensions.

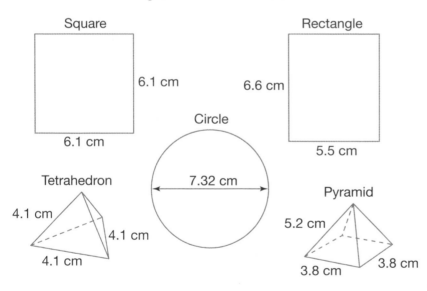

Square

6.1 cm

6.1 cm

Rectangle

6.6 cm

5.5 cm

Circle

7.32 cm

Tetrahedron

4.1 cm

4.1 cm

4.1 cm

Pyramid

5.2 cm

3.8 cm

3.8 cm

Task 1

Each teabag is made from filter paper and filled with dried tea leaves. It is sealed, either with heat or vegetable gum.

1   Three-dimensional teabags, such as the pyramid and tetrahedron, are made from filter paper cut into the nets of their shapes, then filled with tea and sealed. Sketch the net for:

   a a pyramidal teabag

   b a tetrahedral teabag.

2   Flat teabags, such as the square, rectangle and circle, are made from tea leaves sandwiched between two large rolls of filter paper. They are sealed and cut into the teabag shape.

   Use this information and your nets from question **1** to investigate further.

   a Which shape of teabag requires the greatest length of seal?

   b Which shape requires the shortest seal?

Task 2

Brewco, a manufacturer of teabags, buys two rolls of filter paper, each of width 125 mm and length 20 m.

1   a How many of the square teabag shapes can fit across the width of one roll?

   b How many of the square teabag shapes can fit along the length of one roll?

   c Use what you know about how teabags are made, from Task 1, and your answers to **1a** and **1b** to work out how many square teabags can be made from the two rolls of filter paper.

2   a How many of the rectangular teabag shapes can fit across the width of one roll?
      (Think about how you could arrange the rectangles in different ways.)

   b Use what you know about how teabags are made, from Task 1, and your answer to **2a**, to work out how many rectangular teabags can be made from the two rolls of filter paper.

3   A teabag can be made from two circles of diameter 7.32 cm. Two circles of diameter 7.32 cm fit on a square of filter paper, as shown on the right.

   Use what you know about how teabags are made to work out many circular teabags can be made from the two rolls of filter paper.

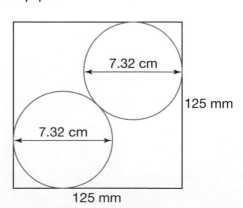

7.32 cm

125 mm

7.32 cm

125 mm

4 Two nets for the tetrahedral teabag fit on a rectangle of filter paper, like this.

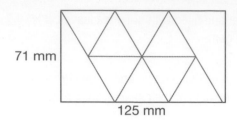

71 mm

125 mm

Use what you already know about how teabags are made to work out how many tetrahedral teabags can be made from the two rolls of filter paper.

5 Two nets of the pyramidal teabag fit on a rectangle of filter paper, like this.

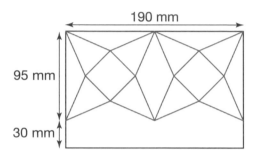

190 mm

95 mm

30 mm

How many pyramidal teabags can be made from the two rolls of filter paper?

Task 3

1 Brewco fills each tea bag with 2.6 g to 3.3 g of tea leaves. Use your calculations from Task 2 to work out how much tea Brewco's factory manager needs for the two rolls of filter paper if:

    a square teabags are filled with 3 g of tea

    b rectangular teabags are filled with 3.3 g of tea

    c circular teabags are filled with 3.2 g of tea

    d tetrahedral teabags are filled with 2.6 g of tea

    e pyramidal teabags are filled with 2.8 g of tea.

    Give your answers in kilograms, correct to 2 decimal places.

2 Brewco pays 0.2 pence per gram of tea leaves. Use your answers to question **1** to work out how much the tea leaves cost for the two rolls of filter paper if Brewco is making:

 a square teabags

 b rectangular teabags

 c circular teabags

 d tetrahedral teabags

 e pyramidal teabags.

  Give your answers in sterling.

3 Including delivery, 50 rolls of filter paper cost £32.50. How much do the two rolls cost Brewco?

Task 4

1 Use your answers to Task 2 and Task 3 to copy and complete these sentences. Replace the first blank with the number of teabags and the second with the cost of filter paper and tea leaves.

 a ___ square teabags can be made from two rolls of filter paper for a total cost of ____ .

 b ___ rectangular teabags can be made from two rolls of filter paper for a total cost of ____ .

 c ___ circular teabags can be made from two rolls of filter paper for a total cost of ____ .

 d ___ tetrahedral teabags can be made from two rolls of filter paper for a total cost of ____ .

 e ___ pyramidal teabags can be made from two rolls of filter paper for a total cost of ____ .

2 Use the sentences above to work out how much one teabag of each shape costs to produce.

Task 5 (extension)

Use the internet or visit a supermarket to conduct a survey of the sale price of teabags of different shapes.

Imagine you are the commercial director for Brewco. Use the information you have collected in your survey, and Tasks 1–4, to write a short report for the board of the company, recommending two shapes of teabag for the company to make. Give reasons for your decisions and use mathematical calculations to support them.

HOW DID YOU FIND THESE TASKS?

- What did you find easy or difficult about these tasks?
- Did you work on your own, in pairs or in groups, and how did this help or hinder your approach and success with these tasks?
- What did you learn about how maths is used and applied in real-world situations?

Small-scale farming

Small-scale farming falls into two categories, **for profit** and **not for profit**. Not for profit farming is usually undertaken by people that have a small piece of land and want to grow their own produce.

The focus of this activity is small-scale farming for profit. Multinational companies are now working on large-scale farming with massive fields and huge herds of animals; this warrants the name of **intensive farming**. Small-scale farmers have to grow their own produce, find their own buyers and sell their goods in very competitive markets.

Learning objectives

Representing Level 1/2: find information needed to solve the problem and use appropriate methods to solve it

Analysing Level 1/2: work through the calculations and check that answers make sense

Interpreting Level 1/2: use results to produce graphs, explain findings and draw conclusions

LINKS WITH
Science
Geography
Business studies
Media studies

Task 1

Robert and Helen live on a small farm in West Yorkshire. They grow potatoes in their largest field, which is a square with sides measuring 140 m.

> 1 acre is equivalent to 4046.86 m^2

1 a How big is the field, in square metres and acres?

   b Seed potatoes should be planted 9 inches apart, with 24 inches between rows. Approximately how many seed potatoes can Robert and Helen plant?

Potatoes are bought and sold by the tonne. There are approximately 3000 potatoes to the tonne.

Seed potatoes cost £100 per tonne. Each seed potato will yield from nine to 15 potatoes.

Potatoes sell for £207 per tonne.

   c What price can Helen and Robert expect their potato crop to fetch? What is their basic profit?

Robert bought fertiliser for the potatoes. He mixed nitrogen (N), phosphorus (P) and potassium (K) in the ratio 6 : 6 : 12. Nitrogen costs £11.50 per litre, plus VAT, phosphorus and potassium cost £22.75 per litre, plus VAT.

2 a How much of each chemical did Robert need to make a fertiliser mixture of 32 litres?

   b Robert dilutes his concentrated fertiliser into a 1000-litre container. If he sprayed his potato crops at a rate of 1 litre per square metre, find the total cost to spray the field.

Helen goes through the accounts after the potatoes have been sold. She has fuel costs of £1070, labour costs of £14 250 and machinery hire of £10 500.

3 Taking all of the costs into account, have Robert and Helen made a profit on their potatoes?

Task 2

Helen keeps 20 dairy cows in a small field all year.
She is paid 24.08 pence per litre by a local dairy.

1 What was Helen's income from the herd last year?

As well as feed, Helen has to pay £940 per month towards the rent, rates and upkeep of the farm.

2 Did Helen make a profit or a loss last year?

HELEN'S NOTEPAD

87% of cows in milk at any one time

7.9 litres of milk per cow per day

28p per day per cow of additional feed

Task 3

Helen rents several fields next to the farm and buys 150 more cows for £98 000. Helen has read that, with food supplements, a cow can produce an average of 26.4 litres per day.

Helen now has 170 milking cows. 87% are in milk at any one time. Each cow costs 38p per day for food and supplements. She has fixed costs of £650 per week, including rental of new equipment.

1 a Work out how many litres of milk she must sell to break even. Assume the amount she receives per litre remains at 24.08p. Give your answer in standard form.

   b How many litres of milk could be produced each year by her herd? Give your answer in standard form.

   c How much profit could Helen make from her herd in one year? Use your answers to parts **a** and **b**.

2 Helen borrows £100 000 per annum, at a flat rate of 5%, from the bank to help with her expansion. She will repay the loan over two years. How does this affect her break-even figure in her first year?

Task 4 (extension)

Each cow must be milked twice a day. Helen wants to buy an automated milking unit. She will let the cows enter from 9 am to 3 pm every day. These formulae can be used to determine how many stalls an automated unit should have.

Milkings per hour = number of milkings needed ÷ shift length (hours)

Total number of stalls = milkings per hour ÷ 4.5

How many stalls does Helen need?

HOW DID YOU FIND THESE TASKS?

- What did you find easy or difficult about these tasks?
- Did you work on your own, in pairs or in groups, and how did this help or hinder your approach and success with these tasks?
- What did you learn about how maths is used and applied in real-world situations?

Coastguard rescue

In *Coastguard search* you investigated the range of incidents that coastguards have to deal with. You identified different search patterns, to cover as large an area as possible in the shortest possible time.

In this activity, you will allocate your helicopter assets and send them out to rescue people involved in incidents around the coast of the UK.

- Data sheet 1 lists the nine sites around the UK from which coastguard rescue helicopters can be launched and provides information about the helicopters available at each one.
- Data sheet 2 is an A3 map of the UK. The nine sites are identified with red circles.
- Data sheet 3 is a list of incidents.

★ ★

Learning objectives

Representing Level 1/2: interpret the situation or problem and identify the mathematical methods needed to solve them

Analysing Level 1/2: use appropriate checking procedures and evaluate their effectiveness at each stage

Interpreting Level 1/2: interpret and communicate solutions to practical problems, drawing simple conclusions and giving explanations

LINKS WITH
English
PHSE
Geography
Economics
Drama

Task 1

Suppose that you have been put in charge of the search and rescue operation. When an emergency call is received, you must respond. It is vital that you respond quickly, to make the rescue in the shortest possible time. You will have to take into account these factors.

- Based on the speed of the helicopter and the distance travelled, you need to find the approximate time it will take to reach the site of the incident.

- Each person will take an average of 5 minutes to be winched on board the rescue helicopter.

- The helicopter will then need to get to the nearest major town that has medical facilities, as marked on the map.

- Helicopters can refuel at any of the RAF or coastguard bases listed on data sheet 1.

A man has got into difficulties in the water off Littlestone-on-Sea, Kent, trying to rescue his dog.

1 Choose the most appropriate base to send a helicopter.

2 Calculate the distance the helicopter will travel to rescue the man and take him to hospital.

3 Calculate the time it takes.

4 Calculate the cost of the rescue.

Look at data sheets 2 and 3.

5 Which do you think will be the **most expensive** rescue? Find the cost of that rescue.

6 Which do you think will be the **least expensive** rescue? Find the cost of that rescue.

Task 2

It is very likely that two or more incidents will occur at the same time.

You will be given details of a number of incidents around the coast, to which you must respond. You must choose the most appropriate response, to rescue people as quickly as possible.

At each incident there are a number of people to be rescued by the helicopters at your disposal. You must allocate the different assets so that they will get to each of the incidents in the shortest time, as well as being able to carry all the people that need to be rescued.

1 Use the map, and its scale, to work out the total distance and time for each rescue and how much it costs altogether.

2 Summarise your incidents for the day and work out the operational limits of the helicopters at your disposal.

  a How far could they go before they had to turn back?

  b Does it depend on how many people need to be rescued?

Task 3 (extension)

Read the article below from the Westmorland Gazette, which describes what happened when Prince William was involved in a helicopter rescue. The emergency call was made to the Liverpool coastguard.

Prince William in Morecambe Bay rescue drama

2:22pm Tuesday
5th October 2010

By **Emma Lidiard Reporter**

PRINCE William carried out his first mission as a rescue pilot in Morecambe Bay – saving a seriously ill gas rig worker.

RAF Valley, Anglesey, where the Prince is based, was called in to airlift a man from a Centrica Energy oil rig after reports a worker had suffered a suspected heart attack.

The Prince, who was part of a four-man helicopter crew, co-piloted a yellow Sea King rescue helicopter to the scene on Saturday, only two weeks after graduating as a search pilot.

Despite strong winds, the crew arrived at the rig –

one of two off Morecambe Bay owned by the energy firm – 24 minutes after the emergency call.

Within minutes the Prince was back in the sky heading to Blackpool Airport where the sick worker, a 52-year-old from Preston, was transferred by land ambulance to Blackpool Victoria Hospital. A spokesman for William said: "Prince William is pleased finally to be able to contribute to the life-saving work of the Search and Rescue Force.

"He is proud, after two years of intense training, to be able to serve in one of Britain's foremost emergency services."

Centrica Energy confirmed one of its contractors had been taken ill on

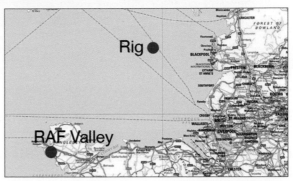

October 2, at around 2pm.

A spokesperson said: "The safety of all our staff and contractors is always our priority and the airlift was carried out smoothly and to plan.

"The support of organisations like the RAF and Coastguard is always appreciated and allows us to operate as safely as possible."

The seriously-ill contractor was taken to the

Lancashire Cardiac Centre at Blackpool's hospital, where a spokesperson said he was in a "comfortable" condition.

The Prince is expected to spend three years as a front-line search and rescue pilot for the RAF working in the North West, which could potentially bring him to South Lakeland, said an RAF spokesman.

Write the script for a role-play between the controller at the RAF headquarters and the pilot of the helicopter. The controller must instruct the pilot where to go, to carry out the rescue, and where to deliver the injured person.
Consider:

- the location of the RAF headquarters
- the location of the place to which the injured person will be taken
- the location of the oil rig
- the distance and bearing between these locations
- the average speed of the helicopter
- the weather conditions.

HOW DID YOU FIND THESE TASKS?

- What did you find easy or difficult about these tasks?
- Did you work on your own, in pairs or in groups, and how did this help or hinder your approach and success with these tasks?
- What did you learn about how maths is used and applied in real-world situations?

Tomorrow's world

The first programmable machine, intended for working out complicated mathematical calculations, was the difference engine, designed by Charles Babbage (1791–1871). It was the size of a wardrobe.

However, electronic computers were not developed until the twentieth century. Those used in the 1960s took up enormous spaces, needed clean, air-conditioned environments and worked with huge reels of magnetic tape.

By today's standards, even the computers of the 1990s were very slow. Storage, on magnetic and floppy disks, was limited and expensive.

What further advances in technology does the next decade have in store?

Task 1

In the UK, most people use **broadband** for connecting to the internet.

A frequent question is: *Am I getting what I pay for, from my broadband provider? How can you tell?*

Speed is measured in **bits per second**, but the numbers would be very large so the more common unit is **megabits per second** (Mbit/s). Essentially, the bigger the number, the faster the connection. Data sheet 1 provides some background information.

1　a　Look at the first table on data sheet 1. Find the actual speed as a percentage of the advertised speed.

　　b　Most people pay for broadband speeds up to 8 to 10 Mbit/s. Produce two comparative bar charts to show the speed over 24 hours and the advertised package speed, maximum and minumum, as shown on data sheet 1. Comment on your chart.

Beverley in East Yorkshire has what may be the slowest broadband in the country. On 17 September 2010, this was put to the test when a 200 megabyte video was uploaded to a video uploading site. As the upload started, a pigeon carrying a memory card with a copy of the video was released and flew to Wrangle in Lincolnshire. The bird arrived after a 75-minute flight.

2 **a** By the time the pigeon had arrived, only 30% of the video had been uploaded. What was the total time that the video took to upload?

b What was the average number of megabytes uploaded per minute?

Task 2

The file size of a music track or video is determined by the quality of the sound or image, as well as its duration. A standard music track is from 3 to 4 megabytes (3–4 MB).

1 **a** Using the information on data sheet 2, calculate the time it will take to download a volume of 32 songs, via both 5 Mbit/s broadband and 56k dial-up. Comment on your answers.

b If you wanted to watch a TV programme live on the internet, which type of connection would you **not** be able to use? Explain why.

c Many internet providers have download limits. Assuming you have a 10-gigabyte download limit per month, how many of each type of entertainment file would you be able to download each month?

2 Your broadband speed depends on many factors, such as the distance from the exchange (or, in some cases, the green telephone company boxes on the streets) to your house, the type of cable between the exchange and your house and the number of people using the same line as you.

a Describe the graph below, which shows the capacity of three different types of line, and explain any key features.

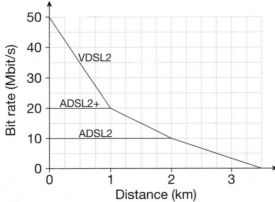

Comparison of FTTH and DSL downstream capacity with distance from exchange

FTTH = Fibre To The Home
DSL = Digital Subscriber Line
(ADSL and VDSL are variants)

b Using the information in this graph, what would you expect your broadband speed to be if you lived 2.5 km away from the exchange?

c Some providers are offering 40 Mbit/s broadband speeds. How close to the exchange would you have to be to receive such a high speed?

Task 3

In the early 1990s everyone used floppy disks as portable devices on which to store files. As file size increased, so did the capacities of portable media.

1   **a** Use an internet search to find the size of typical media storage devices.

     b How many 108 kB text documents could be stored on each type of medium?

     c How much bigger are modern storage devices than floppy disks and other older types? Show your answers as simple ratios.

2   A 10-megapixel photograph occupies approximately 4.7 MB, a downloaded film is around 600 MB and music about 4 MB per track.

     a How many of each type of media files could you store on each of the modern storage devices?

     b How much free space would you have on each device after loading the maximum amount of each type of file?

3   **a** Media players come in various sizes. Find three prices for media players and calculate the cost per gigabyte of capacity for each. Comment on your answers.

     b Estimate the cost of a new model of 128 GB media player, based on your earlier answers.

4   You decide to buy a new 32 GB media player. After formatting, you have 81% free space that you can use. You decide to use 40% for music, 25% for photos and 35% for podcasts.

> Photo: 4.7 MB
>
> Music: 4 MB per track
>
> Podcast: speech only 0.5 MB per minute,
>   audio 1 MB per minute,
>   video 4 MB per minute

     a Calculate the amount of space available for music and photos and find how many of each type of file your media player will store.

Your computer is set up so that an entertainment website will automatically download a new radio broadcast each week (speech) and an educational science video (video). Each programme is 45 minutes long.

     b You transfer these programmes to your media player. How long will it be before it is full?

Task 4 (extension)

How could you 'future-proof' your next computer? Think about much disk space you will need. Consider including an external hard disk to back everything up.

Here are some suggestions for your research.

- Use an internet search, or specialist magazines, to investigate suitable computers.

- Find out about hard drives and external drives, comparing capacities and considering value for money.

- Consider what software you will need, and whether you could use open-source packages.

- Estimate how much space to allow for music, pictures and downloads such as films.

Produce a short report, presentation or documentary to summarise what you have found out and to make recommendations to someone who is considering buying a new computer. Include different options, based on the amount of money they have available.

HOW DID YOU FIND THESE TASKS?

- What did you find easy or difficult about these tasks?
- Did you work on your own, in pairs or in groups, and how did this help or hinder your approach and success with these tasks?
- What did you learn about how maths is used and applied in real-world situations?

Time management

How well do you manage your time? Are you always late? Do other people seem to fit more into their lives than you? Or are you actually one of those people who plan what they want to do and how much time they will spend doing it?

Do you find yourself glued to a social networking site, waiting for someone to post something on their wall so that you can comment, when – instead – you could be out playing sport, doing homework or doing something else you enjoy?

Task 1

Schools in the UK have different starting and finishing times but, whichever school you attend, you should receive the same number of hours of education each year.

Adam goes to a school in North Yorkshire. His school times for the school year 2010–2011 are shown in this table.

Holiday	Schools close	Schools open
Summer break		Wednesday 1 September
Autumn mid-term	Friday 22 October	Monday 1 November
Christmas break	Friday 17 December	Tuesday 4 January
Spring mid-term	Friday 18 February	Monday 28 February
Easter break	Friday 1 April	Monday 18 April
Easter weekend	Thursday 21 April	Tuesday 26 April
May bank holiday	Friday 29 April	Tuesday 3 May
Summer mid-term	Friday 27 May	Monday 6 June
Summer break	Monday 25 July	Thursday 1 September

Adam starts school at 0845. He has registration and then five one-hour lessons each day. School finishes at 1515.

1 a Calculate the proportion of time Adam spends at school during the year 1 September 2010 to 31 August 2011.

   b What percentage of the time that Adam is at school does he spend in class?

   c Think about your own school timetable. What percentage of your total time in classes do you spend on each subject? Check your answers. Comment on your results.

Task 2

During the summer holidays, Adam works at the Scarborough Sealife Centre. His hours are from **10.15 am to 3.30 pm** on Monday, Wednesday and Friday, and from **10 am to 6 pm** on a Saturday.

1 a Adam is paid £3.64 per hour during the week and gets time and a half for weekends. How much does he earn in one week?

   b Adam lives in Muston. It takes Adam 10 minutes to walk from the bus stop to the Sealife Centre. Look at the information in the bus timetable below. Which buses could he catch to get to work on time? Give a reason for your answer.

Bus timetable Monday to Saturday							
Hunmanby	0803	0830	0925	0920	–	1040	1130
Muston	0808	0836	0930	0926	1027	–	1135
Filey	0516	0847	0936	0936	1035	1100	1145
Gristhorpe	0824	0854	0943	0942	1041	–	1150
Cayton	0830	0858	0935	0947	1046	–	1155
Eastfield	0835	0901	–	–	1048	–	–
Wheatcroft	0839	0907	0942	0953	1053	–	1200
Scarborough	0845	0918	0951	1000	1105	1125	1210

2 Express as fractions of the whole summer day, the amount of time Adam spends at work and travelling. Estimate what proportion of the day he spends sleeping and what proportion he spends doing other things.

Task 3

Adam likes to take part in sport. He goes to football training on a Wednesday evening, after school, for 2 hours. He also plays football (90 minutes) for a local team on a Saturday afternoon. He trains at the gym for an hour on two evenings a week and for 2 hours on a Saturday morning.

Adam also has a paper round after school every Monday, Tuesday, Thursday and Friday, which takes him an hour. He has 4 hours of homework set every week. He spends 2 hours a week doing research or coursework and 5 hours a week downloading music and films. His household chores take him about 3 hours a week.

1   Investigate the proportion of time Adam spends on different types of out-of-school commitments. Estimate how much free time Adam has for watching TV and socialising. Produce a chart or graph to display your results and comment on Adam's use of this time.

2   Compare your own use of time to Adam's. Do you have the right balance in your routine?

Task 4 (extension)

1   a   Research effective ways to revise for exams. Make notes of the key points.

   b   Produce a revision timetable for the subjects you are taking, for future exams. Be realistic in your planning or you will not stick to it. Allow additional time for research and any trips to the library you may need.

   c   What proportion of your time have you left for work, sleep and socialising?

HOW DID YOU FIND THESE TASKS?

- What did you find easy or difficult about these tasks?
- Did you work on your own, in pairs or in groups, and how did this help or hinder your approach and success with these tasks?
- What did you learn about how maths is used and applied in real-world situations?

GM foods

Genetic modification, or genetic engineering, occurs when scientists change the genetic material (DNA) of an organism, which may be a plant or an animal, in a way that would not happen naturally. For example, some plants have been genetically modified to make them resistant to disease or make the fruit ripen more slowly, so they can be in transported and stored for longer than normal, without going bad. Genetically modified foods, known as **GM foods**, currently consist mainly of crops such as soybeans, corn, maize and cotton-seed oil.

The first plant-based GM foods were put on the market in the early 1990s, although they were not approved in the UK. By July 2010 products from GM animals had been developed but not yet put on the market.

Task 1

Officially, there are currently no GM foods being grown in the UK since they have not been commercially approved, although some have been grown experimentally, under close supervision. According to released statistics, most of the worlds GM crops are grown in the USA (46%), Brazil (16%), Argentina (15%), India (6%), Canada (6%), China (3%), Paraguay (2%) and South Africa. In this task you will compare the production of GM crops in industrial countries and developing countries.

1   a   What percentage of GM crops are grown in South Africa and the other countries not included in the list above?

    b   What percentage of GM crops are grown in industrialised countries and what percentage are grown in the other countries of the world, including developing countries? You may find the data sheet helpful when you answer this question.

2 The table shows the numbers of hectares (in millions) set aside for growing GM crops in industrial and developing countries in 2004 and 2009.

	2004	2009
Developing countries	25.3	57.3
Industrial countries	51.6	72

a What was the total amount of land used for GM crops in 2004 and in 2009?

b Use your answer to part **a** to find the percentage of land used for GM in developing countries and in industrialised countries, relative to all of the land that is used for GM crop production, for both 2004 and 2009. Give your answers correct to one decimal place.

c Brazil is considered to be a developing country. How could the information you have found in part **b** be used to show this?

d Use your answer to part **b** to predict how many hectares will be used for GM crops in developed and industrial countries in 2014. Check your answer.

Task 2

There are four main types of crop in which GM varieties are important. The areas of land used for these crops, in both GM and conventional forms, are shown in the bar chart. The **yield** is the amount that can be harvested from a plant each year. GM crops have been modified to produce high yields to feed the growing populations.

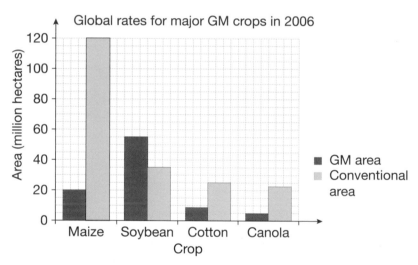

1 a Use the information in the bar chart to find the percentage of each crop grown worldwide that was GM modified.

b What was the global percentage of GM crops grown in 2006?

c Approximately what fraction of each crop (maize, soybean, cotton and canola) was a GM crop? Show each fraction in its simplest form.

2 a The total area of the four crops in the bar chart is 88.1 million hectares. Write this number in standard form.

> 1 hectare is equivalent to 0.1 square kilometres
> 1 hectare is equivalent to 3.86×10^{-3} square miles

b What is an area of 88.1 million hectares in square kilometres and in square miles?

c The area of the British Isles is 84 300 square miles. How many times bigger was the area used for GM crops in 2006, compared to the whole of the British Isles?

Task 3

Large amounts of GM crops were produced in 2006. Since then, not only has the amount of these crops increased, but other types of food are now GM modified. In 2008, 125 million hectares of farmland were used to grow GM crops.

Are people concerned about what they are eating and what affect it might have on their health? The results of a survey of over 1000 people, carried out in 2010, to find out what people thought of GM foods are shown in this graph.

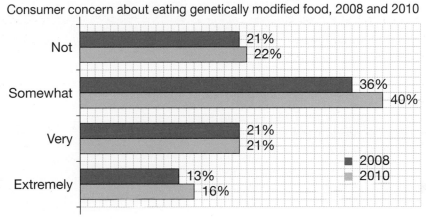

Consumer concern about eating genetically modified food, 2008 and 2010

1 What conclusions can you draw from the results of this survey?

2 In 2010, the population of the UK was approximately 61 million. Is this survey a good representation of the population? Give reasons for your answer.

Task 4 (extension)

1 **a** In pairs, use the internet to find information about GM foods.

b Write a **hypothesis** about GM foods and produce a questionnaire to test it. Design a data-collection table to record people's opinions. Remember to use closed questions.

2 Conduct your survey, asking at least 20 people.

3 **a** Think about the most appropriate ways to present your data, for example, in charts, tables or diagrams.

b Draw simple conclusions for any charts you produce and write a statement to explain what they show.

c Have the results of your questionnaire proved your hypothesis? Comment on your findings and how accurate your hypothesis was.

HOW DID YOU FIND THESE TASKS?

- What did you find easy or difficult about these tasks?
- Did you work on your own, in pairs or in groups, and how did this help or hinder your approach and success with these tasks?
- What did you learn about how maths is used and applied in real-world situations?

Flying the world

At any time, as the Sun is shining brightly on some parts of the Earth, other parts are in shadow and it is night. Imagine if the whole world used the same time. Then noon in some places would be the middle of the day, but in other places it would be mid-afternoon, or late in the evening. To avoid this, we have **time zones**, so that we can be sure that, wherever we are in the world, noon always occurs in the middle of the day and approximately when the Sun is highest in the sky.

This means that when you fly to somewhere abroad, the time you arrive depends not only on the distance you travel but also on the time zone of the country to which you are travelling.

Learning objectives

Representing Level 1/2: undertake problem-solving in an unfamiliar context and identify the necessary information to tackle the problem

Analysing Level 1/2: apply relevant mathematics to find solutions to practical problems

Interpreting Level 1/2: interpret results to multistage practical problems and then draw appropriate conclusions

LINKS WITH
Geography
English
Science

Task 1

Look at the data sheet, which shows the Earth divided into 24 different time zones, based roughly on the lines of longitude. These are imaginary lines on the Earth's surface, passing through the North and South poles.

1   Longitude 0° passes through Greenwich, in England, and lies in the middle of the time zone. Times in the first time zone to the East of 0° are +1 hour, compared to the time in Greenwich, and those in the first time zone to the West of 0° are −1 hour.

Mark all time zones on the map (top and bottom).

2   Copy the table below.

Departure city	Arrival city	Time difference (hours)	Length of flight (hours)
Paris	Los Angeles		
Tel Aviv	Lima		
Lisbon	Hong Kong		
Sao Paulo	Tokyo		
Auckland	Rome		

a Use the time zones on the map to help you work out the time differences between the departure and arrival cities.

b Use the time differences and the departure and arrival times scheduled below, to help you work out the duration, or length of time, of flights between the cities.

Departure city	Departure time (local time)	Arrival city	Arrival time (local time)
Paris	12:30 Tuesday	Los Angeles	15:00 Tuesday
Tel Aviv	10:00 Wednesday	Lima	18:20 Wednesday
Lisbon	13:45 Monday	Hong Kong	11:00 Tuesday
Sao Paulo	08:20 Friday	Tokyo	20:20 Saturday
Auckland	07:15 Sunday	Rome	21:00 Sunday

Task 2 (extension)

Use the time zones on the map, and the schedule below, to match the three pairs of people to one of these flights to holiday destinations, leaving on Saturday morning.

Departure city	Departure time (local time)	Arrival city	Arrival time (local time)
London, UK	11:20 Saturday	Vancouver	13:10 Saturday
London, UK	10:00 Saturday	Cape Town	23:25 Saturday
London, UK	10:40 Saturday	Dubai	20:40 Saturday
London, UK	11:25 Saturday	Banjul	16:10 Saturday

Husband and wife Jonathan and Sam have the following requirements.

- Jonathan would like to travel business class but doesn't want to spend too much money, so is looking for a flight of less than 6 hours.

- Sam would like to finish reading her book and needs a flight lasting more than 2 hours.

Sisters Saskia and Zinnia have the following requirements.

- Saskia suffers with jet lag and would like the time difference between the UK and their holiday destination to be 2 hours or less.

- Zinnia likes to start her holiday with a long flight so she really feels like she is getting away. She would like a flight lasting more than 7 hours.

Friends Max and Zac have the following requirements.

- Max has three movies he wants to watch on the flight, so would like it to be at least 6 hours long.

- Zac can get bored on long flights and would ideally like the flight to take no longer than 8 hours.

Task 3

Remember the D, S, T triangle.

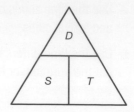

$$\text{speed} = \frac{\text{distance}}{\text{time}} \qquad \text{time} = \frac{\text{distance}}{\text{speed}} \qquad \text{distance} = \text{speed} \times \text{time}$$

1   Copy and complete the table below. You will need to use the length of flights you calculated in Task 1, question **2b**, with the correct formula, to work out the average speed for each journey. Give your answers correct to 1 decimal place (1dp).

Departure city	Arrival city	Distance (km)	Speed (km/h)
Paris	Los Angeles	9029	
Tel Aviv	Lima	12685	
Lisbon	Hong Kong	11044	
Sao Paulo	Tokyo	18560	
Auckland	Rome	18270	

2   Use the average speeds you found in question **1** to estimate the average speed of an aeroplane.

3   Copy and complete the table below. Use the time zones on the map on data sheet 1 and the distances from question **1** to help you estimate the distances. Then use your answer to question **2**, with the correct formula, to estimate the lengths of flights. Give your answers correct to the nearest hour.

Departure city	Arrival city	Estimated distance (km)	Estimated length of flight (hours)
Taipei	Madrid		
Brisbane	Reykjavik		
San Francisco	Brussels		
Buenos Aires	Shanghai		
Santiago	Addis Ababa		

Task 4

Most planes do not fly in straight lines. Instead, they each follow a flight path to avoid other aeroplanes or unsafe areas. However, the more direct their route, then the more they can save on fuel costs.

Here are the details of the flight paths for two of the journeys from Tasks 1 and 3.

Lisbon → Rome (bearing _____) → Dubai (bearing _____) → Hong Kong (bearing _____)

Paris → Reykjavik (bearing _____) → San Francisco (bearing _____) → Los Angeles (bearing _____)

Draw these flight paths on the data sheet.

Now copy and complete the paths above, measuring and writing down the bearing for each part of the journey.

Task 5

Here is a three-day roster for a member of a flight cabin crew.

SUNDAY 1 SEPTEMBER

    10:45 LOCAL TIME, REPORT IN AT LONDON, UK

    FLIGHT NUMBER CE029   12:15 LOCAL TIME, DEPART LONDON, UK

                                        20:15 LOCAL TIME, ARRIVE DUBAI, UAE

MONDAY 2 SEPTEMBER

    09:50 LOCAL TIME, REPORT IN AT DUBAI, UAE

    FLIGHT NUMBER CE038   11:20 LOCAL TIME, DEPART DUBAI, UAE

                                        1335 LOCAL TIME, ARRIVE ADDIS ABABA, ETHIOPIA

  REPORT AT 22:15 LOCAL TIME, ADDIS ABABA, ETHIOPIA

    FLIGHT NUMBER CE041   23:45 LOCAL TIME, DEPART ADDIS ABABA, ETHIOPIA

TUESDAY 3 SEPTEMBER

                                        05:15 LOCAL TIME, ARRIVE LONDON, UK

  06:45 LOCAL TIME, LONDON UK, CLEAR TIME

Imagine you are responsible for planning rosters. Use the data sheet, your knowledge of time zones and length of flights to create your own three-day roster for a member of flight cabin crew. (You may make up flight numbers.) You must follow these three rules.

Rule 1: Cabin crew must report for a shift 1.5 hours before a flight takes off.

Rule 2: Cabin crew shifts do not end until 1.5 hours after a flight lands.

Rule 3: The minimum rest period between shifts depends on the lengths of flights, as shown in the table.

Flight type	Flight time	Minimum rest period
Short-haul	< 3 hours	1 hour
Medium-haul	3 – 6 hours	5 hours
Long-haul	> 6 hours	11 hours

Once you have created your roster, answer the qustions below.

1   Work out the length of time for each flight.

2   Estimate the total distance, in kilometres, the cabin crew travels in the three days.

3   Use the data sheet to draw a flight path for each part of the journey and work out the bearings.

HOW DID YOU FIND THESE TASKS?

- What did you find easy or difficult about these tasks?
- Did you work on your own, in pairs or in groups, and how did this help or hinder your approach and success with these tasks?
- What did you learn about how maths is used and applied in real-world situations?

Your plaice or mine?

The UK has always had a thriving fishing industry. Fishing ports, such as Hull, have grown and many local people have prospered from the fishing trade.

Now that EU laws are limiting the amount of fish that can be caught, people are asking, 'What is the fate of the UK fishing industry and the economy that depends on it?'

To complicate things, what were once British territorial waters are now being exposed to exploitation from other EU member countries. In spite of a recent recovery, reserves of cod in the North Sea have dropped to 3% below the previous healthy levels. There is a real risk that they will fall to unsustainable levels. Most cod now sold in Britain is imported from Iceland.

Learning objectives

Representing Level 1/2: find required information from graphs and charts to solve the problem, using appropriate methods

Analysing Level 1/2: work with scales on charts and diagrams and ensure that calculations make sense

Interpreting Level 1/2: use results to explain findings and draw conclusions

LINKS WITH
Science
Geography
English

Task 1

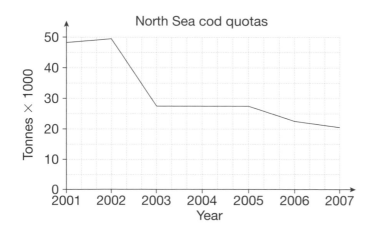

North Sea cod quotas

1 a Study the graph and describe the trend that it shows. Estimate the cod quota for 2008.

   b Find the reduction in the cod quota from 2001 to 2007.

   c Express the cod quota for 2007 as a fraction of the 2001 quota. Approximate your answer and reduce the fraction to its simplest form.

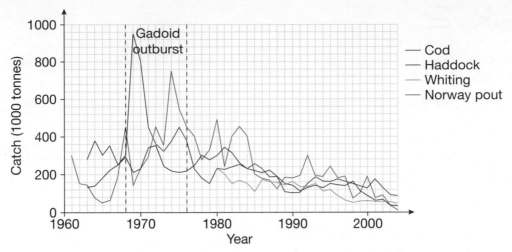

2 This graph shows the quantities of the four most common types of fish caught in the North Sea.

The **gadoid outburst** was a short period when fish such as cod, haddock, whiting and Norway pout became abundant in the sea for no apparent reason. This resulted in the cost per tonne falling dramatically. It has not been repeated.

a What does the graph show about fishing catches in the North Sea?

b What happened to the catches of fish during the gadoid outburst period? Give specific examples.

c Which type of fish has consistently yielded the best catch?

d Which type of fish was not recorded as being caught during the late 1960s or the 1970s? Explain how you know this.

3 The UK has two main limits imposed on the number of fish it can catch. The EU sets a limit, called a **quota**, on the amount of fish that can be caught. The UK has to make do with a share of this quota.

a Use this bar chart to estimate the percentage of fish the UK was given as its EU quota, out of the total stock for each year.

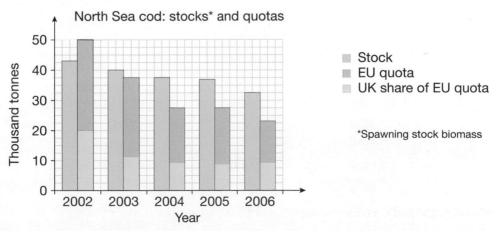

b Calculate the remaining stock levels of fish for each year. Give reasons for the decrease in stock levels.

Task 2

1   Use these two line graphs to comment on the employment level in the fishing industry between 1997 and 2006.

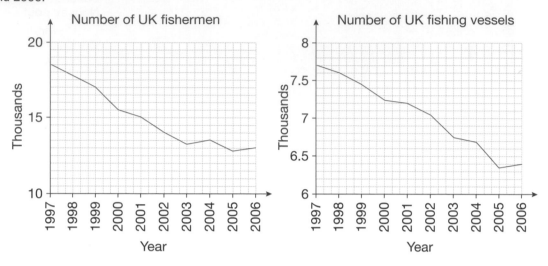

2   What happened to the number of fishermen per vessel from 1997 to 2006?

Task 3

The table shows the cost, in euros, of different types of fish over four different years.

Average fish price (euros per tonne) by fish species and year				
	1991	1997	2000	2002
Cod	1406	1433	2073	2269
Dogfish	660	605	617	1370
Haddock	1242	983	1166	1344
Hake	3102	3365	3214	3440
Monk/angler	2439	2306	3005	3226
Plaice	1305	1768	2237	2653
Ray/skate	829	842	1225	1110
Mackerel	189	439	410	541
Tuna	3624	1857	2125	1861
Crab	929	992	1234	1340
Crawfish	21 159	17 448	17 908	24 040
Lobster	9551	11 037	13 837	12 916
Shrimp	6297	5916	6800	6905
Squid	2048	2506	1608	2080

1   Use the graph to find out how many tonnes of haddock, cod and plaice were caught in 1997, then calculate how much money this brought into the UK that year.

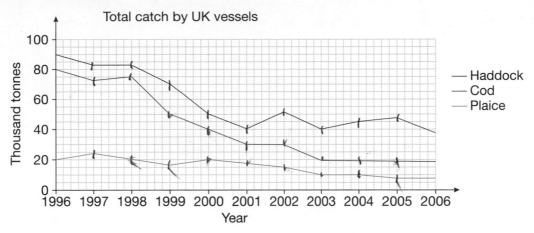

2   How does this compare to the amount of money brought into the UK for these fish in 2000 and 2002?

Task 4

The map and pie chart show fishing areas around the UK and the percentage catch by sea area.

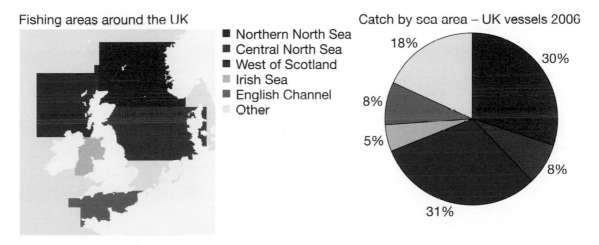

What does the information in the map and the pie chart demonstrate?

HOW DID YOU FIND THESE TASKS?

- What did you find easy or difficult about these tasks?
- Did you work on your own, in pairs or in groups, and how did this help or hinder your approach and success with these tasks?
- What did you learn about how maths is used and applied in real-world situations?

Sleep

Every animal, including humans, needs sleep to stay healthy. While we sleep our bodies restore our energy levels. Too little sleep can leave us feeling grumpy; too much can leave us feeling listless.

There are two main types of sleep. The first, **NREM** (non rapid eye movement) sleep, has four stages, ranging from light to deep sleep. The second, called **REM** (rapid eye movement) sleep is characterised by bursts of rapid eye movement while the eyes remain closed. This doesn't mean the eyes are constantly moving, but they dart back and forth or up and down. They stop for a while and then start again.

During sleep, people experience repeated **cycles** of NREM and REM sleep. First is the NREM phase, lasting approximately 90 to 110 minutes. This is repeated four to six times per night, interspersed with periods of REM sleep. As the night progresses, the amount of deep NREM sleep decreases and the amount of REM sleep increases.

Learning objectives

Representing Level 1/2: understand problems in familiar and unfamiliar situations, and collect and represent data, using ICT where appropriate

Analysing Level 1/2: use probability to assess the likelihood of an outcome

Interpreting Level 1/2: extract and interpret information from tables, diagrams, charts and graphs

LINKS WITH
Science

The amount of sleep we need varies from person to person and changes with age. Research suggests that regularly missing just one hour of sleep a night can affect our performance, thinking abilities and mood.

While muscle tone is normal during NREM sleep, we are almost completely paralysed during REM sleep. Some people occasionally experience panic attacks when they semi-wake from REM sleep and cannot move. Although the muscles that move our bodies go limp, other important muscles continue to work during REM sleep. These include the heart, diaphragm, eye muscles and smooth muscles such as those of the intestines and blood vessels.

Task 1

The graph shows a typical sleep pattern, illustrating a normal cycle.

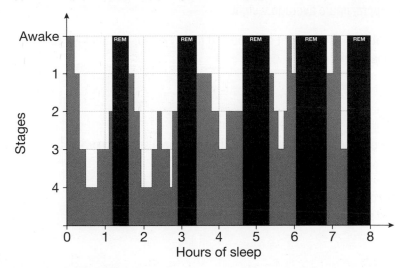

We are much more likely to remember our dreams if we are awoken from REM sleep. Some people think that this is because REM sleep is the time when we dream. Others believe we dream throughout our sleep, but only remember them if awoken from REM sleep.

Assuming we can wake up at any stage, use the chart to calculate the probability that we wake during REM sleep and consequently remember our dreams.

Task 2

The amount of sleep required for the body and brain to function well varies from person to person. The bar chart shows the average amount of sleep needed at various ages, and the percentage of REM sleep that occurs.

Use the bar charts to calculate the amount of REM sleep, in hours, experienced per night by each age group. Present your data in the form of a table.

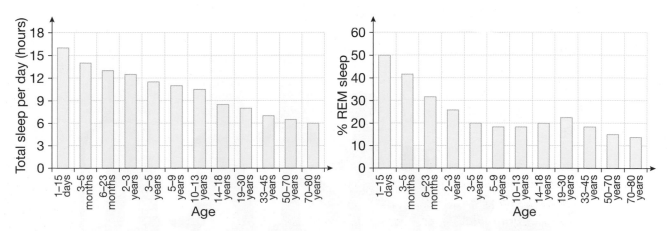

Task 3

Different animals also need different amounts of sleep. The data sheet shows the amounts of sleep needed for various species, and an approximate average weight.

Is there any correlation between the amount of sleep and the weight of the animal? Using a spreadsheet might make the work more manageable.

Task 4 (extension)

The list of animals on the data sheet is so varied that it might make sense to group them into different categories.

1   Draw graphs to find out if there is a correlation between weight and amount of sleep for:

   a   cats (C)

   b   rodents (R)

   c   primates and humans (P)

   d   ruminant mammals (M).

2   Describe the correlation in each case.

Task 5

1   Carry out a survey to investigate the relationship between people's ages and the hours of sleep they usually get. You will need to design a data collection sheet and then survey as many people as you can, across different age ranges.

2   Present your findings in a graph, chart or table.

3   How do your findings compare to those in the bar chart in Task 2?

Task 6

The onset of puberty in humans brings with it a change to the biological clock, which creates a tendency to go to bed later and wake later.

- Sleep requirements for adolescents do not decline as they mature. In fact, there is evidence that they need more sleep, as they grow. Despite this, adolescents generally sleep less than they need to, due to increased socialising and school work.

- While adolescents require between 8.5 and 9.25 hours of sleep per night, over 25% report getting less than 6.5 hours of sleep on school nights.

1 Carry out a survey among adolescents to find out if they are getting sufficient sleep.

2 Find out the lesson times during the school day when they feel most and least awake.

3 Analyse your results to decide whether there might be evidence to support changing the times of the school day.

HOW DID YOU FIND THESE TASKS?

- What did you find easy or difficult about these tasks?
- Did you work on your own, in pairs or in groups, and how did this help or hinder your approach and success with these tasks?
- What did you learn about how maths is used and applied in real-world situations?

Selling online

You may think that selling on the internet is good news for businesses. They don't need to pay rent on high-street shops, nor keep changing their shop window display; they can be open 24 hours per day, seven days per week. However, there are still many costs involved.

Task 1

Online sales as a percentage of total retail sales

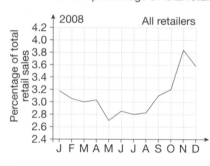

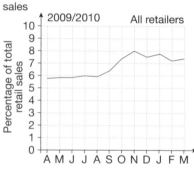

These graphs are adapted from the UK Office of National Statistics.

1   Look at the horizontal axis on each graph. What do you notice about the time of year each of them starts?

2   Look at the vertical axes. What do you notice about the scales used?

3   Do the graphs show that internet sales are rising or falling annually?

4   What are the internet sales as a percentage of retail sales in:

   a November 2008   b November 2009?

5   Why do you think there is a small peak in November, in both 2008 and 2009?

6   Give two reasons why you think the graphs show a small dip in internet sales as a percentage of retail sales in December in 2008 and 2009.

Task 2

Pooch Palace

Book your shampoo and groom online and make your dog happy for just £20. Click here to make your payment now. Use code: woof

Flying high

Send us your design and we'll make you a personalised kite – £15 is all you'll pay!

Hats off to this deal!

Choose your hat colour and your bobble and we'll do the rest. Keep warm in style this winter for only £8.49 incl pp!

1 Copy and complete this table for each of the online businesses advertising above.

	January internet sales	February internet sales	March internet sales	Total number of sales
Pooch Palace	87	96	72	
Flying high	61	89	132	
Hats off	204	329	124	

Most online payments are made with a debit or credit card. For every payment that is made, the seller pays a fee. This ensures the card payment is secure and reaches their bank account safely. The fee is either a monthly flat rate and then a payment for each sale or it is a percentage of every sale.

Deal A: £15 flat rate per month plus 28p for every sale.

2 Use the adverts above and your completed table to work out the cost over the three months of Deal A for:

   a *Pooch Palace*   **b** *Flying high*   **c** *Hats off*.

Deal B: 4% of sale price

3 Work out the cost of Deal B for a single sale for:

   a *Pooch Palace*   **b** *Flying high*   **c** *Hats off*.

4 Use your completed table above and your answer to question **2** to work out the cost over the three months of Deal B for:

   a *Pooch Palace*   **b** *Flying high*   **c** *Hats off*.

5 Which deal would you recommend for:

   a *Pooch Palace*   **b** *Flying high*   **c** *Hats off*?

Task 3

Pay-per-click is a type of internet advertising. A business places an advert on someone else's web page. When a potential shopper clicks on the advert they are taken to the business's website. The business pays every time a shopper clicks through, regardless of whether they buy anything. Therefore, the business must be very careful not to spend too much on each click!

> Online businesses use this calculation to find out the cost of the advert:
>
> **cost of advert = cost per click × number of clicks**
>
> Then they use this calculation to find the profit they have made from the advert:
>
> **profit = (sale price − cost price) × number of sales**

They want the cost of the advert to be less than the profit they have made, otherwise pay-per-click advertising doesn't make good business sense.

Pooch Palace, Flying high and *Hats off* all decide to try pay-per-click. Look at the adverts in Task 2 and the data below. Show the calculations that support your answer to each question.

1. *Pooch Palace* place their online advert on a large pet store's website. They agree to pay £0.09 per click and in the first month they get 5765 clicks.

    For each shampoo and groom it costs them £12 for materials and labour. This is the **cost price**.

    They make 53 sales from the online advert. Should *Pooch Palace* continue with this advert?

2. *Flying high* place their online advert on a large toy store's website. They agree to pay £0.11 per click and in the first month they get 4012 clicks.

    For each personalised kite it costs them £4.50 for materials and labour. This is the cost price.

    They make 94 sales from the online advert. Should *Flying high* continue with this advert?

3. *Hats off* place their online advert on a winter holiday website. They agree to pay £0.15 per click and in the first month they get 10 924 clicks.

    For each bobble hat it costs them £1.83 for materials and labour. This is the cost price.

    They make 257 sales from the online advert. Should *Hats off* continue with this advert?

Task 4

Vincent's Usability Lab conducted research into the most noticeable online adverts. They set up a computer showing 15 web pages, in random order. Each web page contained news, articles and a single advert.

Each advert was placed in a different position on the web page (top left, top middle, top right, in a box in the middle or across the bottom) and had a different design (text only, text and picture of product, text and picture of a person using the product).

132 people took part in the experiment. They were asked to click through each of the 15 web pages, reading as much or as little as they wanted to. When they had finished they were asked which advert was most memorable. The results are shown in the table.

	Design		
Position	Text only	Text and picture of product or service	Text and picture of person using product or service
Top left	7	17	11
Top middle	2	18	30
Top right	6	15	19
In a box in the middle	0	2	4
Across the bottom	0	0	1

1 What is the probability of a person noticing an advert placed:

  a top left   **b** top middle   **c** top right   **d** in a box in the middle

  e across the bottom?

2 Where is: **a** the best place   **b** the worst place to put an advert?

3 What is the probability of a person noticing an advert with:

  a text only   **b** text and pictures of products   **c** people using products?

4 What is: **a** the best design   **b** the worst design for an advert?

Task 5 (extension)

Use your answers to Task 4 to write a short paragraph analysing the design of the online adverts in Task 2.

Task 6

Fancy that sells all its fancy dress costumes for £12.50 each. It makes an average profit of £3.50 on each one.

Imagine that, as an online business consultant, you have been asked to advise *Fancy that* on selling online. Use what you have learnt in Tasks 1 to 5 to write a short report offering your advice. Include:

- why you advise them to sell online
- which deal you advise them to go with initially for card payments, and why
- any tips you have for advertising on the internet.

HOW DID YOU FIND THESE TASKS?

- What did you find easy or difficult about these tasks?
- Did you work on your own, in pairs or in groups, and how did this help or hinder your approach and success with these tasks?
- What did you learn about how maths is used and applied in real-world situations?

Jewellery design

Have you ever thought about becoming a maker of jewellery?

Nearly everybody buys jewellery at some point in their lives. Necklaces, rings and items for body piercings are everyday items. Jewellery makes a very practical birthday or Christmas present.

There are three stages that jewellery goes through before anyone can wear it:

- design
- manufacture
- retailing.

Some people specialise in one of these areas, but many others in the industry design, make and sell the jewellery themselves.

Areas of work in which jewellers are involved include:

- designing and making jewellery, which also covers fashion jewellery
- goldsmithing or silversmithing
- production benchwork, which involves mounting and setting precious and semi-precious gem stones and resizing jewellery
- selling jewellery to customers and advising them, for example, about valuation, repair and alteration services.

Task 1

Jewellers refer to diamonds and other precious gems as stones. Most raw diamonds have dull, rough external surfaces. The diamond-cutter's art is in polishing the stone, creating flat faces in a symmetrical arrangement to make it shine and sparkle.

Learning objectives

Representing Level 1/2: understand, use and calculate ratio and proportion, including problems involving scale

Analysing Level 1/2: solve problems requiring calculation with common measures, including money, time and length; use a formula

Interpreting Level 1/2: construct geometric diagrams; extract and interpret information from tables, diagrams, charts and graphs; interpret data and draw conclusions

LINKS WITH
Art
Design and technology
Science
Business

The diamond has to be cut so that most of the light shining on the surface is reflected, like the beam from a torch shone at a mirror. The angle of each cut is crucial in making a diamond sparkle brightly.

Too shallow   Ideal   Too deep

Point cut   Table cut   Old single cut   Mazarin cut   Peruzzi cut   Old European cut

    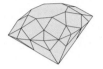

The diagrams above show how the shape of cut stones has changed over the years, as design and cutting techniques have developed.

This diagram shows the plan view of diamond cut in the Old European style.

A jewellery designer might start with a drawing like the one in the diagram on the right.

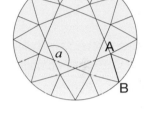

1 Reproduce the design by following the instructions on data sheet 1.

2 **On your drawing**, measure the angle marked a and the length of AB in the diagram.

3 If a diamond has a diameter of 8 mm, what will be the size of angle a and the length of AB?

Task 2

When you make, buy or sell a ring, you need to make sure it is the right size. Data sheet 2 gives details of ring sizes for the UK, the USA and Europe. Can you help these three people to work out their correct ring sizes?

1 An American woman, who knows her ring size is 9, wants to buy a ring in the UK. What UK sizes are equivalent to a USA size 9?

2 A French man wants to buy a ring. In France, he is a size 62.

   a What size would he need if he bought a ring in the USA?

   b What size might he need if he bought a ring in the UK?

3 A woman measured her finger by wrapping a piece of string around it, marking the overlap with a pen and measuring the length between the marks on the string. It was 4.8 cm. What size ring should she buy in the UK?

Task 3

If you are interested in making jewellery, you could start by doing it as a hobby, making inexpensive costume jewellery.

You would need to buy a pair of pliers, to tighten up the fasteners, and then any other materials you wanted to use.

The table shows some basic materials you could use to make some simple jewellery.

Number	Item	Notes	Price
1 set	Three pairs of pliers	For cutting wire, tightening fixings	£9.95
1	Large pendant bead	For necklace	£1.50
100 g	Plain glass beads	Enough for necklace 75 cm long	£1.50
5 g	Opaque Czech beads	Enough for necklace 75 cm long	£0.60
100	Long ball wires	For earrings ('fish hooks')	£1.00
100	Jump rings 5 mm	For one end of necklace	£1.00
1	Trigger clasp	Clasp with trigger catch for other end of necklace	£0.10
1 m	Heavy chain	For necklace, could attach large pendant	£2.50
10 m	Plastic coated wire	Silver coloured	£0.60

All necklaces will need a jump ring and a trigger clasp.

1   How much would it cost to make:

   a   a heavy 1-metre chain necklace with a large pendant bead

   b   a 75-cm necklace of plain glass beads on plastic-coated wire

   c   a 75-cm necklace with a half-and-half mixture of plain glass beads and opaque Czech beads, on plastic-coated wire

   d   a pair of 'fish-hook' earrings with a large pendant bead on each?

2   Now suppose you are going to sell each item you have made. You want to make some profit because you spent time making them, and you had to spend at least £9.95 on the pliers. Suggest a reasonable selling price for each item.

Task 4 (extension)

Charlie decides to set up in business as a jeweller, using a spare room at home as a workshop.

The table shows the costs and the time it takes for Charlie to complete some tasks.

Task	Materials	Wholesale cost (£)	Time (hours)
Resize a ring	None	None	1.5
Make a standard ring from a stock ring and a diamond	Silver ring	6	2
	Gold ring	30	
	Diamond (small)	108	
	Diamond (medium)	212	
	Diamond (large)	432	
Make a ring to customer's design	Silver	4	8
	Gold	21	

Charlie adds 30% to the cost of any materials and charges £30 per hour for the time spent working on the job.

Below is a list of Charlie's transactions during one month.

Charlie estimates that the overhead costs (heating, lighting, rates) are £11 000 per year.

Assuming the month described is typical for the year, what is Charlie's annual profit after paying for overheads?

Type of job	Quantity
Resized rings	5
Made and sold rings:	
Silver, small diamond	7
Silver, medium diamond	3
Silver, large diamond	1
Gold, small diamond	2
Gold, medium diamond	7
Gold, large diamond	4
Rings made to customer's design:	
Gold	1
Silver with three small diamonds	1
Gold with large diamond	1

HOW DID YOU FIND THESE TASKS?

- What did you find easy or difficult about these tasks?
- Did you work on your own, in pairs or in groups, and how did this help or hinder your approach and success with these tasks?
- What did you learn about how maths is used and applied in real-world situations?

Buying your first car

One of the most useful qualifications you will ever earn is your driving licence. Without a car you have to rely on public transport so you will have to work close to where you live. Your social activities will be limited.

Having passed your test, what car would you choose? Would you prefer one with a special image or something that may be reliable, if not very exciting? Remember that the bigger the engine, the faster it can go, but also the more it will cost in fuel, insurance and road tax.

Buying a car is the easy part. Running, insuring and maintaining it can be rather expensive.

Cars that are three or more years old must have a valid **MOT** to be driven legally on public roads.

The car must also be **insured** before it can be driven on public roads. This can be expensive, but becomes cheaper as you become older and more experienced and don't have any accidents.

These **fixed costs** have to be paid however far you drive. Then there is **maintenance**, which includes regular services, usually every 20 000 miles, brakes, new tyres, exhausts and anything else that can break, corrode or fall off your car.

Finally there is **fuel**. The good news is that the less you use your car the less fuel you have to buy. This is a **variable cost**.

Task 1

What would you like as your first car?

1   List **five** cars you would consider buying as your first car.

2   In groups, combine your lists and find the top two cars. You must be able to justify why they could be sensible choices as a first car.

3   In your group, estimate how much it will cost to run your top two cars for the first year, including insurance for an 18-year-old.

4   Use information provided by your teacher to check how accurate your estimates were.

Task 2

1   Using the internet, your local newspaper or car magazines, find 10 suitable first cars. Design your data collection table first.

2   Use the information you have found to work out the mode, mean, median and range for price. Find the mean and range for mileage.

3   Using your table, draw suitable bar charts to compare price and mileage. You could use a spreadsheet program to draw a simple diagram to compare both sets of data on the same graph. Comment on your findings.

Task 3

Cars are put into road-tax bands depending on how much they pollute the air. The further down the alphabet, the more you have to pay.

Road tax banding			
Band	CO_2 emission (g/km)	Rate for 12 months (£)	Rate for 6 months (£)
A	Up to 100	0	0
B	101–110	35.00	NA
C	111–120	35.00	NA
D	121–130	120.00	66.00
E	131–140	120.00	66.00
F	141–150	125.00	68.75
G	151–165	150.00	82.50
H	166–175	175.00	96.25
I	176–185	175.00	96.25
J	186–200	215.00	118.25
K	201–225	215.00	118.25
L	226–255	405.00	222.75
M	Over 255	405.00	222.75

Mia decides to buy a Vauxhall Corsa 1.2 litre Club (petrol model). The Vauxhall Corsa is in road-tax band G.

The car has done 36 000 miles and is for sale for £2970. Mia is told that it does 46 mpg, which means that, on average, it will travel 46 miles on one gallon of fuel. Insurance on this car for an 18-year-old female is £1195 per year.

1 Using all the information you have, work out the cost of buying and running the car for the first year, not including maintenance and fuel.

UNLEADED
pence per litre — 120.0

DIESEL
pence per litre — 126.0

2 The average person drives 12 000 miles in a year. How much will Mia spend on fuel if she drives 12 000 miles during the first year? 1 gallon is equivalent to 4.55 litres.

3 Car fuel tanks vary in size. Use the conversion graph to calculate what it will cost to fill:
a a 10-gallon tank with petrol
b a 17-gallon tank with petrol.

4 An annual MOT and service will cost £250, if booked together. Maintenance will cost a further £150 a year.

Using your answers to questions **1** and **2**, calculate the total cost for buying and running Mia's car for the first year.

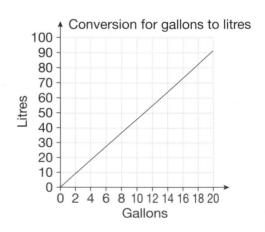

Conversion for gallons to litres

Task 4

Buying a new car can save servicing costs, as some manufactures offer three-year warranties and free servicing, and the MOT isn't required for the first three years. On average, a new car loses 40% of its initial value in the first year, then depreciates at a steady 10% per year.

The new Smart car is in road-tax band A

1 If you buy a new Smart car for £9995 how much will it be worth after five years?

2 Many people cannot afford to pay the full price for a new car, so they buy it on **finance**. This means that they pay **monthly instalments** (an amount of money each month) over several years, but they also have to pay **interest** as a percentage of what they owe. If Tom bought a Smart car on finance, at a flat **interest rate** of 9% on the purchase price over five years, how much would he repay per month?

3 If he wanted to sell his Smart car after three years, would he be able to pay off the **remaining finance** with the money he gets from the sale of the car? Comment on your answer.

4 If Tom decides he wants a bigger car and sells his Smart car after the first year, will he be able to pay off the remaining finance with the money he makes from selling the car? Comment on your answer.

Task 5 (extension)

Using the internet, research a car you would like to own from new.

1 Calculate the annual running costs of your car. (You might not be able to include insurance.)

2 How much money would you lose if you were to sell the car after two years?

HOW DID YOU FIND THESE TASKS?

- What did you find easy or difficult about these tasks?
- Did you work on your own, in pairs or in groups, and how did this help or hinder your approach and success with these tasks?
- What did you learn about how maths is used and applied in real-world situations?

Fish and chips

For many years, fish and chips was the only readily available takeaway meal in the UK. Despite the increase in takeaways, for example, Chinese food, Indian food, Thai food and home pizza deliveries, fish and chips is still among the most popular takeaway meals.

In fact, fish and chips is so popular in Britain that fish and chip shops have opened in other countries, such as Spain, to cater for British holidaymakers.

Task 1

Refer to the data sheet to answer these questions. Write the answers in ordinary numbers and in standard form.

1   How many fish and chip shops are there in the UK and Ireland?

2   How many fish and chip shops were there in the 1930s?

3   How many portions of fish and chips are sold every year?

4   A typical portion of fish and chips weighs 220 g.
    How much fat would this contain?

5   What is the total weight of fish and chips eaten every year in the UK and Ireland?

6   If a burger and medium fries weighs 331 g, how much fat would this contain?

Task 2

There is a danger that supplies of cod will run out. If the stocks get too low, fish will not be able to breed in sufficient quantities to replace the number being caught.

Use the data sheet to answer these questions.

1   Which is the most popular type of fish used for fish and chips?

2   Of the three types of fish, cod, haddock and herring, which has had the lowest stock in 1975?

3   Roughly what was the mass of the cod stock in 1975?

4   Which fish had higher stocks in 2005 than in 1963?

5   The haddock stock in 1970 was 900 000 tonnes, but it had fallen to 185 000 tonnes by 1979. What was the percentage decrease in stock?

Task 3

The data sheet shows prices of cod and haddock from a wholesale fish market in a fishing port.

1   Which fish is cheaper for the fish and chip shop to buy, cod or haddock?

2   Why do you think it is cheaper?

3   A medium fish (cod or haddock) might weigh about 400 g before it is gutted and filleted. How much does one fish cost the fish and chip shop owner?

4   A portion of chips might contain about 100 g of potato. The fish and chip shop owner can buy potatoes at 30p per kilogram. How much does a medium cod and a portion of chips cost the owner?

5   The price in the shop for cod and chips is £2.50. What is the percentage profit to the owner?

6   What other costs does the owner have?

Task 4 (extension)

A fish and chip shop uses 60 litres of cooking oil a week. This creates 20–25 litres of waste oil, which it has to dispose of. In London alone, more than 50 million litres of used oil are produced each year.

Businesses have to pay to have the used oil taken away and disposed of at landfill sites. This costs the shop about £15 each time it is collected. A typical shop would need to have it collected every fortnight.

There is a new scheme that turns fat into fuel. The fat is put through a revolutionary new process, which transforms it into a form of fuel called **biodiesel**, which can be used in cars. It is environmentally friendly and most diesel cars will run on it.

An organisation in Norfolk produces over 200 000 litres of biodiesel a week, using oil from restaurants, takeaways and crisp factories.

There is no waste at all from this process, as the waste oil is turned into biodiesel, glycerine and fatty acids, which can be used in central heating oil.

It gives off less harmful emissions than ordinary diesel, and has less impact on air quality.

Crude oil?

1   How much could the owner of a fish and chip shop hope to save in a year by giving used oil to a biodiesel producer?

2   Use the internet to find the advantages and disadvantages of biodiesel compared to normal (petrol) diesel.

Task 5

Read the newspaper article opposite, then answer these questions using the figures given in the article.

1   a How much, in percentage terms, are the fish and chip shop owners saving by selling fish varieties other than cod or haddock?

   b If a fish and chip shop owner sells 250 portions (400 g per portion) of 'scam' fish in one day, how much extra money is he or she making compared to when selling cod or haddock?

2   a Do you think the fish and chip shop owners should risk getting caught and having to pay the fine? Provide reasons.

   b Think about and discuss the long-term problems that can be caused by selling fish that is contaminated with bacteria and industrial waste.

Great fish swindle is scam and chips

By **Daniel Jones and Clare Kane**

REAL

FAKE

PROFIT: Cod, top, is twice the price of pollock, bottom

CHEATING chippies are up to something fishy – up to a QUARTER of them are flogging cheapo varieties as high-priced cod.

Unsuspecting customers are being served catfish, farmed in polluted Asian rivers, costing as little as half as much.

Sold at the same price as fresh cod or haddock, the CON and chips scam is netting the unscrupulous owners big profits.

But it could be at an even bigger cost to the health of British families.

Some of the fish has been found to be contaminated with bacteria and industrial waste, which could cause long-term problems.

Two recent reports, by the Food Standards Agency and Dublin University, show that between one-tenth and one-quarter of chippies are conning customers this way.

One of the most common fish passed off as cod is Pangasius, one of 20 types of catfish farmed in cages along Vietnam's heavily-polluted Mekong Delta.

Not as flaky as cod and with a blander taste, the white fish's wholesale price is around £5 per kg compared with up to £11.75 per kg for the real thing.

Pollock, coley and whiting are also being used as cheaper substitutes.

A number of chip shop owners have already been fined up to £6000 each for the rip-off.

Phil Whitehouse, from Worcestershire Trading Standards, said: 'It's probably a much bigger problem than we realise.'

So watch out: It could be Lying Tonight at your local chippy!

The Sun 12/09/10

HOW DID YOU FIND THESE TASKS?

- What did you find easy or difficult about these tasks?
- Did you work on your own, in pairs or in groups, and how did this help or hinder your approach and success with these tasks?
- What did you learn about how maths is used and applied in real-world situations?

The Milky Way

A **galaxy** is a collection of billions of stars. Our sun is a star and our galaxy is called the **Milky Way**. It is believed to be about 13 billion years old – almost as old as the **Universe** itself. The Universe contains billions of galaxies. No one knows exactly how many there are, as most of them are simply too far away to see, even with today's powerful telescopes.

A **light year** is the **distance** that light travels in a year. The Milky Way has a diameter of about 100 thousand light years and contains between 200 and 400 billion stars. With numbers this big, who needs to be exact!

Each star you see in the night sky is actually a sun and may have planets orbiting it. Our Sun has nine planets, including Pluto; some have none and others have a few hundred.

Learning objectives

Representing Level 2: identify information needed from a table and choose from a range of mathematics to solve problems

Analysing Level 2: apply mathematics in an organised way to find solutions, and use appropriate checking procedures

Interpreting Level 2: explain what results show, draw conclusions and explain findings pictorially, as appropriate

LINKS WITH

Science

ICT

History

English

Task 1

Astronomers, who study the stars, have calculated distances between the objects in our galaxy. They have done this by measuring the angle at which we observe them, from two separate positions, and using geometry.

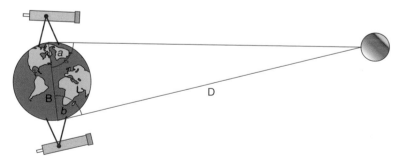

1 How close to Earth is our nearest neighbour? How distant is the planet furthest away from Earth?

2 a Using the table on data sheet 1, write down masses of the planets, in standard form.

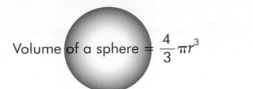

$$\text{Volume of a sphere} = \frac{4}{3}\pi r^3 \qquad \text{Density} = \frac{mass}{volume}$$

b Using the known masses and diameters of the planets on the data sheet, calculate the density of each planet.

c Which planet(s) could have a similar composition to that of Earth?

Another way to describe the size of a planet is to compare it to the mass of the Earth.

d Using the values for mass of each planet, compare how many times bigger or smaller than Earth the other planets are. Give your answers correct to one decimal place.

Task 2

On 21 July 1969, Neil Armstrong became the first man to walk on the Moon, leaving behind his famous footprints, in the Sea of Tranquillity. Less famously, and only 100 feet away, he left a two-foot panel containing 100 mirrors, all facing the Earth.

By aiming a laser beam at the Moon, and measuring the time it takes for the light to travel to the Moon and back, we can calculate the distance to the Moon from the Earth.

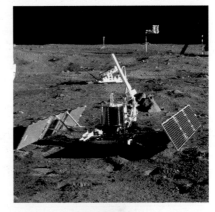

1 If the speed of light is 3×10^8 metres per second (m/s) and it takes 2.58 seconds for the light to reach the Moon and return to the Earth, how far is the Moon from the Earth?

2 How long does it take the light from the Sun to reach us?

$$\text{speed} = \frac{distance}{time}$$

Task 3

The planets in our solar system orbit the Sun. Their orbits are not exactly circular, because the gravity of the other planets pulls them in different directions.

When new and very distant stars are discovered, they are rarely visible. Their presence is deduced because of evidence of their gravitational pull on the paths of objects that we can actually see.

1 a Using a suitable scale, and the information from data sheet 1, draw the orbits of the planets around the Sun. Label each orbit and represent the planets with circles of appropriate sizes.

b What are the maximum and minimum distances between the Earth and Saturn in their orbits?

Task 4

Nothing moves faster than light, which travels at 300 000 000 metres per second (m/s) or, in standard form, at 3×10^8 m/s. In comparison, sound travels much more slowly, at approximately 330 m/s, which is a mere 3.3×10^2 m/s.

The electromagnetic spectrum

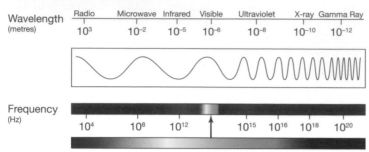

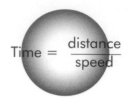

$$Time = \frac{distance}{speed}$$

1 **a** Approximately how much greater is the speed of light than the speed of sound?

   b If lightning strikes a tree one mile away (1.6 km), how long will it take the light from the strike to travel to your eye? Give your answer in standard form, and as an ordinary number.

2 **a** How long will it take the sound of the tree being struck to reach your ear?

   b Suppose that you see lightning in the sky and hear the crash of thunder 17 seconds later. How far away is the storm in miles?

Dr Choudhury, who works for NASA, has sent a space probe (Robbie) to Pluto to explore the surface. The landscape is uneven, with some canyons over half a mile deep. Dr Choudhury's probe is radio controlled. Radio waves, as part of the **electromagnetic spectrum**, travel at approximately the speed of light.

Dr Choudhury activates Robbie when the distance of the orbits brings Mars to **7.8×10^{10} m** from the Earth.

3   **a** How long will it take the command for Robbie to move forwards to reach him?

Robbie is moving at a speed of 0.5 m/s. On the screen, Dr Choudhury sees the probe heading for a cliff!

   b If the cliff is 250 m away from the probe's position, as shown on the screen, will Dr Choudhury's command for stop reach Robbie in time?

   c Comment on your answer and compare it to those of your classmates.

Task 5 (extension)

The NASA space shuttle travels at about 17 500 miles per hour, or 7800 metres per second (m/s).

Investigate the possibilities of space travel to the planets in the Milky Way at this speed and at the speed of light.

HOW DID YOU FIND THESE TASKS?

- What did you find easy or difficult about these tasks?
- Did you work on your own, in pairs or in groups, and how did this help or hinder your approach and success with these tasks?
- What did you learn about how maths is used and applied in real-world situations?

On your bike

The earliest record of a bicycle dates back to the 1490s, when a student of the artist Leonardo da Vinci sketched a vehicle that looks remarkably like the bicycles we ride today. However, it wasn't until 1865 that the first two-wheeled riding machine came into popular use. It was called the **velocipede** (Latin for 'fast foot') or, more commonly, the boneshaker, because its wooden structure and metal tyres made it rather uncomfortable to ride on cobbled streets! Since then, metal frames, gears, pneumatic tyres and many more innovations have been adopted in bicycle design.

TRICYCLE.
EARLY FORMS OF CYCLE

The "Dandy-Horse."

3. Early Bicycle with cranks.

4. Gompertz's Velocipede.

7. The "Ordinary" Bicycle.

5. The Dublin Velocipede.

6. The "Bone-shaker."

8. The "Rover" Safety (original form)

118

Task 1

The 'high wheeler' or 'ordinary' bicycle was introduced in 1870. It had one very large front wheel (with diameter ranging from 42 inches to 60 inches) and one small rear wheel (with diameter ranging from 14 inches to 20 inches).

This bicycle was later nicknamed the **penny-farthing** after the large old English penny (with diameter 30.81 mm) and the much smaller farthing (with diameter 20 mm).

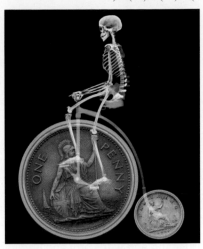

1   Use mathematics to work out if the ratio between the wheels of the penny-farthing bicycle is ever representative of the ratio between the coins it is named after. Show your working, and write your conclusion.

2   The penny and the farthing coins in the name of the bicycle are examples of the 'old money' used in Great Britain until 1971. Back then, there were 240 old pennies in a pound. The farthing, worth a quarter of an old penny, was taken out of circulation at the end of 1960. On 15 February 1971 a new decimal system of coins was introduced, with 100 new pennies in the pound.

a   Use the table below to determine if there have been any decimal coins since 1971 that better represent the ratios between the wheels of the penny-farthing.

Decimal coin	Diameter (mm)
Halfpenny (1971–1984)	17.14
1 penny	20.32
2 pence	25.9
5 pence	18
Old 5 pence (1968–1989)	23.59
10 pence	24.5
Old 10 pence (1968–1992)	28.5
25 pence (1972–1981)	38.61
1 pound	22.5
2 pounds	28.4
5 pounds	38.61

b   What nickname might you give the bicycle?

3 On a penny-farthing bicycle, the pedals are attached to the front wheel only. This means when you turn the pedal through one revolution, the front wheel goes through one revolution too.

a Write an equation for the distance (s) the penny-farthing moves forward in terms of the diameter (d) of the large wheel with one turn of the pedal.

b Write an equation for the distance (s) the penny-farthing moves forward in terms of the diameter (d) of the large wheel with the number of turns (p) of the pedals.

c Rearrange your equation for **b** so that it gives the number of turns (p) of the pedals in terms of the diameter (d) of the large wheel and the distance (s) travelled.

Task 2 (extension)

Husband and wife Larry and Geraldine own two penny-farthings. On a two-mile cycle ride Larry counts that he turns his pedals 747 times. Geraldine has worked out that, for every turn of her pedals, she moves forward 9.4 inches less than Larry does for one turn of his pedals. Use the equations you wrote for Task 1, question **3** to help you calculate the following.

1 The diameter of the large wheel on Larry's penny-farthing, to the nearest inch.

2 The diameter of the large wheel on Geraldine's penny-farthing, to the nearest inch.

3 How many more times Geraldine turns her pedals than Larry does on their two-mile cycle ride.

Task 3

Today's bicycles tend to have a pair of pedals in the middle that drive one or more large gear wheels at the front end of the chain. When the pedals are turned, the chain pulls round one of several small gear wheels attached to the back wheel.

Gear wheels are made to different sizes, depending on the number of teeth.

The gear ratio is:

> number of teeth on the front gear wheel : number of teeth on the rear gear wheel

It is usually written as $n : 1$.

Therefore, gear ratio measures the number of times the rear wheel turns for each revolution of the pedals.

A bicycle with two wheels of diameter 26 inches has front chain wheels ranging from 22 to 42 teeth and rear gears ranging from 12 teeth to 30 teeth.

1   Calculate the gear ratios for the highest gear, which is the greatest ratio; and the lowest gear, which is the smallest ratio.

2   Calculate how far the bicycle moves forward in the highest gear compared to the lowest gear for one revolution of the pedals. Give your answer in inches.

3   The bicycle is ridden at a rate of 60 revolutions of the pedal per minute. What is the range of speed for the bicycle? Give your answer in miles per hour.

   Your answer should demonstrate that the gears enable the rider to climb a steep hill very slowly (in the lowest gear) or race fast (in the highest gear).

Task 4 (extension)

Bicycle A:	Bicycle B:	Bicycle C:
Wheel diameter: 28 inches	Wheel diameter: 22 inches	Wheel diameter: 26 inches
Front chain wheels: From 26 to 56	Front chain wheels: From 24 to 30	Front chain wheels: From 20 to 46
Rear chain wheels: From 10 to 32	Rear chain wheels: From 14 to 38	Rear chain wheels: From 15 to 28
Bicycle weight: 8 kg	Bicycle weight: 11 kg	Bicycle weight: 13 kg
Bicycle tyre width: 20 mm	Bicycle tyre width: 32 mm	Bicycle tyre width: 25 mm

Look at the specifications for the three bicycles. Find the range of speeds for each bicycle if it is ridden at a pedal rate of 60 revolutions per minute. Then use your answer to determine which bicycle would best match each of these activities.

Activity 1: Cycling holiday in a hilly part of Cornwall

Activity 2: Road race on a flat cycle trail around a lake at Rutland Water

Activity 3: Bicycle to go to school across town along bridle paths and cycle paths in Chipping Norton

Task 5

As well as gears, bicycle tyre pressure also affects the rider's speed. It is important to get the pressure right for the width of the bicycle tyre, the load the bicycle is carrying and the type of surface being ridden on. Data sheet 2 shows the *minimum* inflation pressures for six of the most popular tyre widths and for different loads on a wheel (in kilograms). The rider should never let the tyre pressure be lower than this.

How to use data sheet 2

- Find the total load (including the weight of rider, bicycle and luggage). If cycling on poor roads or gravel tracks, then add 25% to the load weight as a safety factor.
- Estimate the load on each wheel. For a normal two-wheeled bicycle carrying one person, assume that 65% of the weight is carried by the back tyre; 35% of the weight is carried by the front tyre.

  This would be different for a tandem, where the ratio may be nearer to 50 : 50.

  Use the loads and the correct tyre–width graph or its equation to work out the minimum tyre pressure for the rear tyre and the front tyre.

1   Find the equation for each of the tyre width graphs, where x represents the load and y represents the tyre pressure.

2   Use the correct equations to work out the recommended minimum tyre pressure for back and front wheels for each of the bicycles A, B, C from Task 4 if:

  a   bicycle A is carrying a rider of 85 kg and no luggage

  b   bicycle B is carrying a rider of 68 kg and luggage of 10 kg

  c   bicycle C is carrying a rider of 59 kg, luggage of 4 kg and goes along some gravel paths.

  Give your answers to the nearest whole number psi, and check them using the graph.

Task 6

A cycle shop has taken on some new weekend staff and you are responsible for planning their training. Use what you have learnt in this activity to write a training manual. You may include:

- questions you would advise the sales staff to ask customers in order to help them choose the right bicycle

- supporting mathematical calculations to help staff work out any relevant information about the bicycles being sold.

HOW DID YOU FIND THESE TASKS?

- What did you find easy or difficult about these tasks?
- Did you work on your own, in pairs or in groups, and how did this help or hinder your approach and success with these tasks?
- What did you learn about how maths is used and applied in real-world situations?

Water usage

Everyone knows that water is a precious commodity that we need to use as carefully as we can. For this reason, showers are becoming ever more popular, as they use much less water than a filling a bath.

In this activity, you will take on the role of someone running a small company that refits bathrooms. To design a functional bathroom you need to take into account factors such as how many people will use it, how fast the water flows and how much hot water you will need. One of your customers has a small bathroom that she wants to have refitted with a shower instead of a bath.

Task 1

A power shower gives a greater flow rate, as it has a pump between the shower head and the hot-water tank, rather than allowing the water to flow through the shower under gravity.

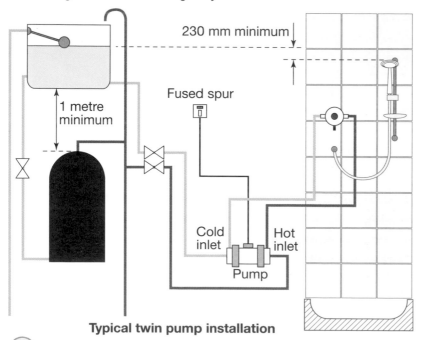

Typical twin pump installation

Your local plumbers' merchant has 28 different types of pump in stock. These are classified by the flow rate and the amount, in litres, of hot water they supply per minute.

Pump flow rate (litres/minute)	Number of pumps
0 ≤ rate < 5	4
5 ≤ rate < 6	6
6 ≤ rate < 7	9
7 ≤ rate < 8	5
8 ≤ rate < 9	3
10 ≤ rate	1

1 a Produce a cumulative frequency graph from the information in the table.

   b Use your cumulative frequency graph to produce a box plot.

   c What is the median flow rate of the pumps available from the plumbers' merchant?

2 a From the information in the table, calculate an estimate of the mean flow rate.

   b How does this value compare to your median?

Task 2

The pipes supplying water, both hot and cold, to showers and baths are wider than those used for basins and sinks. The larger pipes give a higher flow rate; for example, a bath tap will fill a litre bottle more quickly than a tap of a hand basin would.

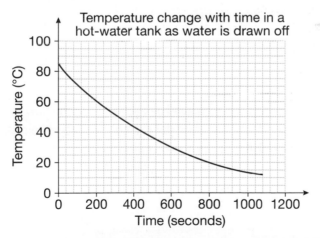

1 a Assume that the temperature in the tank starts at 70 °C. For how long can the shower be run before the water temperature reaches 35 °C?

   b Using the mean shower flow rate that you worked out in Task 1, question **2a**, how much water does a shower that lasts the time you found in part **b** use?

It is important to make sure the customer has a large enough hot-water tank fitted, to supply the water for the house, especially if you are fitting a power shower.

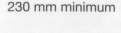
230 mm minimum

$$1 \text{ litre } = 1000 \text{ cm}^2$$

70 cm

50 cm

2　**a** Assuming the tank is full to the top, how much water does the customer's tank hold?

　　b Using the mean shower flow rate that you worked out in Task 1, question **2a**, how many minutes will it take to empty the water tank?

　　c If the mains pressure refills the water system at a rate of 8 litres per minute, will the tank ever be completely empty?

3 How does your mean shower flow rate from Task 1, question **2a** compare with the UK published figure in this table?

Sanitary fittings	Loading value	Flow rate (litres/second)
WC cistern	2	0.1
$\frac{1}{2}$-inch washbasin tap	1.5	0.15
Spray mixer tap	–	0.04
$\frac{3}{4}$-inch bath tap	10	0.3
Shower mixer	3	0.1
$\frac{1}{2}$-inch sink or washing machine tap	3	0.2

Task 3 (extension)

Estimate or record of how much water you use on average each day. Use the flow rates from Task 2 to help you. Investigate water usage around the world (for example, you could compare average water usage in the UK and in Africa or look at why water usage needs to be reduced and how this could be done). Summarise your findings in a report or presentation.

HOW DID YOU FIND THESE TASKS?

- What did you find easy or difficult about these tasks?
- Did you work on your own, in pairs or in groups, and how did this help or hinder your approach and success with these tasks?
- What did you learn about how maths is used and applied in real-world situations?

Crash investigation

Accidents happen unexpectedly. You can't plan for them. However, you can do your best to avoid them. One obvious way is not to take risks when driving.

There are legal speed limits on the roads, with penalties for anyone caught speeding. Are the police justified in enforcing speed limits? Have their activities led to fewer road accidents?

Task 1

Over the past 40 years there has been a significant increase in the number of vehicles on our roads. However, over this time, the number of accidents has decreased.

1 The graph shows the numbers of deaths on British roads, 1967–2003. Can you identify a trend?
  Use numerical data to explain your answer.

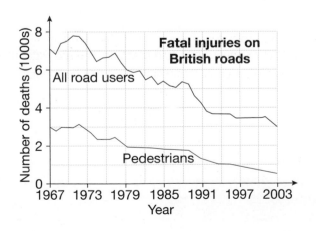

2 **a** Use the first table on the data sheet to compare the chances of being involved in a fatal accident for each type of transport in each year. Which is the safest form of transport?

b Now use the second table on the data sheet to say which form of transport appears to be the safest. How can you explain this, with reference to table 1?

3 This cumulative frequency diagram shows the probability of death for a pedestrian colliding with a vehicle at various speeds.

a Using the graph, what is the probability that a pedestrian involved in a collision will be killed, if the car is travelling at 30 mph, 40 mph and 60 mph?

Probability of death after an accident

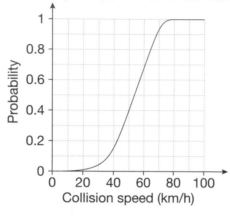

1 mile is equivalent to 1.6 km

b Explain why local councils have changed the speed limits in populated areas from 30 mph to 20 mph.

When there are major roadworks on a motorway average speed cameras are put up to encourage people to drive more slowly. If you drive through an area with average speed cameras, you will see drivers slowing down as they approach the camera and then speeding up once they have passed it.

Average speed cameras work by calculating a vehicle's speed between two points that are a known distance apart. When a vehicle enters a speed-check zone, a camera reads the number plate. The photograph is marked with the date and time. When the same vehicle leaves the zone another camera reads the number plate and, again, the photograph is marked with the date and time. A computer uses the information to calculate the average speed at which the vehicle travelled between the two cameras. If this exceeds the limit, the details are sent to the police for possible prosecution.

Task 2

Andy is driving down the M1 and enters an area of roadworks. Vehicles' speeds are monitored by two cameras, at the start and end of the roadworks. Andy notices his speed at both these points and at eight points in between, as shown in the table. The speed limit is 50 mph.

	Point 1 Site of entrance camera	Point 2	Point 3	Point 4	Point 5	Point 6	Point 7	Point 8	Point 9	Point 10 Site of exit camera
Speed (mph)	50	50	50	45	50	55	57	55	50	45

The distance between the entrance and exit cameras was 9 miles. The difference in time between the photographs at the entrance and exit cameras was 9 minutes 30 seconds.

Andy does not believe he should get a speeding ticket. Is he right? Explain your answer.

Task 3

You are going to take on the role of an accident investigator, to determine the speed of a vehicle involved in an accident. A Peugeot 106 was travelling along a road, in dry conditions, when a pedestrian walked out from behind a stationary ice-cream van.

For the purpose of this investigation, assume that the stopping distance of a car is directly proportional to the square of its speed.

1   **a** Write this relationship algebraically, where D is the stopping distance and v is the speed. Rewrite it as a formula, using k to represent the constant of proportionality.

In test conditions, you applied the brakes, locking the wheels, from a speed of 30 mph. You did this several times. The average length of the skid mark was 33 feet and 4 inches or 33.33 feet.

    b Use this information to find the constant of proportionality.

The speed limit was 30 mph. The police measured the skid marks left by the car to be 50 feet.

2   Use your formula from **1a** to work out how fast the car was travelling.

3   Investigate stopping distances from stationary to 100 mph. Use a suitable method to display your result. Comment on braking distances at 70 mph and above.

HOW DID YOU FIND THESE TASKS?

- What did you find easy or difficult about these tasks?
- Did you work on your own, in pairs or in groups, and how did this help or hinder your approach and success with these tasks?
- What did you learn about how maths is used and applied in real-world situations?

Glastonbury Festival

The first Glastonbury Festival was held in 1970. Since then it has grown enormously. It is now the world's largest open-air festival.

Over the years, despite admission prices steadily increasing, audience figures have rocketed. Therefore, the festival organisers have had to plan for huge numbers of people. The organisers need to provide enough space for campers to set up tents in a safe environment. They have to make sure that food and drink, toilet facilities, water for drinking and washing and medical cover are available. They also need to cooperate with the local police. As well as all this, they also need to make sure that they can cover their costs.

Task 1

The table shows the admission prices to the Glastonbury Festival since 1970.

Year	Price (£)	Year	Price (£)	Year	Price (£)
1970	1	1987	21	2000	87
1971	free	1989	28	2002	97
1978	free	1990	38	2003	105
1979	5	1992	49	2004	112
1981	8	1993	58	2005	125
1982	8	1994	59	2006	133
1983	12	1995	65	2007	145
1984	13	1997	75	2008	155
1985	16	1998	80	2009	175
1986	17	1999	83		

1. What was the percentage increase in admission price from:

   a 1982 to 1983   b 1984 to 1985   c 1989 to 1990   d 2008 to 2009?

2. Investigate how the percentage increase in the admissions price has changed.

3. The ticket price for 2011 is £195. This is a 5.4% increase on the 2010 price.

   What was the price in 2010?

Task 2

Tickets for 85 000 tents are sold for the Glastonbury Festival.

The *Event Safety Guide* recommends a maximum of 430 two-person tents per hectare.

The camping area at Glastonbury is 285 acres.

$$1 \text{ acre} = 4047 \text{ m}^2$$
$$1 \text{ hectare} = 10\,000 \text{ m}^2$$

Below are details of four two-person tents. The floor dimensions are given in centimetres.

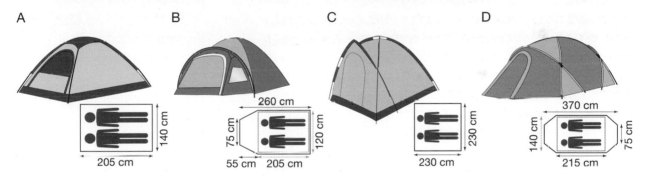

A   B   C   D

The tents at Glastonbury must be pitched so that they do not touch each other. People must be able to walk between them.

1. Estimate how much land area an average two-person tent will need and compare your answer to the allowance given by the *Event Safety Guide* and to the space allowed at Glastonbury.

2. What might make your calculations inaccurate?

Task 3 (extension)

1   Estimate how much standing area each person at the festival needs to view an act in safety.

2   The Pyramid Stage arena at Glastonbury can hold in excess of 90 000 people. How big an area do you think is needed for that number of people?

3   The shape of the viewing area makes a big difference to how well people can see the stage. Draw diagrams of different shaped viewing areas and add lines of sight for people standing at the corners and at the edges of the areas. What shape do you think provides the best view for most people?

Task 4

The following table shows the amount of rainfall, in millimetres (mm), in the Glastonbury area each June since the festival started.

Year	Rainfall (mm)	Year	Rainfall (mm)	Year	Rainfall (mm)
1970	57.6	1984	30	1998	156
1971	130	1985	111.1	1999	83.2
1972	114.1	1986	70.2	2000	51.5
1973	48	1987	103.7	2001	42.6
1974	72.2	1988	45.2	2002	59.7
1975	12.4	1989	46.7	2003	64
1976	20.9	1990	94	2004	64.4
1977	79.5	1991	118.1	2005	78.7
1978	53	1992	34.8	2006	37.6
1979	39.1	1993	96.7	2007	146.7
1980	133.9	1994	41.1	2008	61.4
1981	54.6	1995	19.7	2009	63.4
1982	123.9	1996	37.6	2010	35.9
1983	46.6	1997	125.1		

Here is the same data shown on a graph.

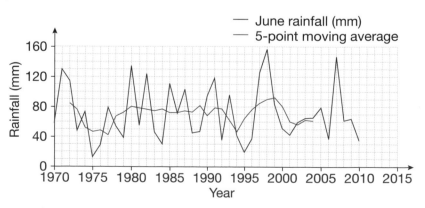

1   Calculate the five-point moving averages, starting with 2003 to 2007 and up to 2006 to 2010.

2   Plot those averages on the graph.

3   Why is a moving average useful?

4   Is there a trend developing in the amount of rainfall in June?

Task 5

Use the information from Tasks 1 and 2 to answer these questions.

1   Looking at increases in price, what might someone wanting a ticket to go to the Glastonbury festival in 2030 expect to pay?

2   The trend in tent design is towards tipi tents.

Use these dimensions to calculate the floor area of a tipi tent.

260 cm

260 cm

3   However, the average number of occupants amongst campers using tipi tents is five. The average tipi tent can accommodate six people.

a   Compare the floor area and the number of campers of a standard tent from Task 2 to a tipi.

b   If everyone camped in tipi tents by 2030, would there be room for more or fewer campers? What is the percentage increase or decrease?

HOW DID YOU FIND THESE TASKS?

- What did you find easy or difficult about these tasks?
- Did you work on your own, in pairs or in groups, and how did this help or hinder your approach and success with these tasks?
- What did you learn about how maths is used and applied in real-world situations?

Leaving smaller footprints

In 1991, the average amount of power used in the UK, per person, was equivalent to a constant 3.6 kilowatts (kW).

Over time the quality of our lives has generally improved, but this has caused an increase in our energy consumption. We use washing machines and televisions frequently, and many people have dishwashers, ipods and computers, all of which use up valuable energy.

The way the majority of our energy is generated depends on the country where we live. Some forms of generating energy are more efficient than others. Most emit carbon dioxide (CO_2).

As shown in this pie chart, transport accounts for the largest end-user of energy. In 1960, it only accounted for 17%. Domestic use (home) has remained about the same since the 1960s.

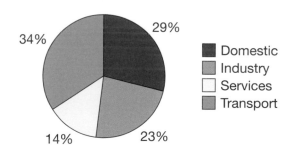

34% 29% 14% 23%

- ■ Domestic
- ■ Industry
- □ Services
- ■ Transport

Energy consumption by final users, UK 2000

Nuclear power station

Coal-fired power station

Wind farm

Task 1

Efficiency is a measure of the amount of **usable** power produced, compared to the total power consumed to generate that power. The table shows the efficiencies of different methods of producing electricity in the UK.

Source of generation	Percentage efficiency (%)
Hydroelectric (large)	95
Tidal power	90
Hydroelectric (small)	90
Coal-fired plant	45
Oil-fired plant	44
Gas turbine	39
Nuclear fission	36
Biomass	35
Municipal waste	25
Solar thermal	18

1   Use appropriate mathematical methods to compare the efficiencies of the different types of power station given in the table. Comment on the usefulness of any averages you use. Choose a suitable chart to display your findings clearly.

2   Coal-fired power stations have been used for many years. Comment on their suitability as power sources, making reference to your findings from question **1**.

3   The diagram shows the efficiency of a coal-fired power station.

This can be written as:

energy input = waste energy output + useful energy output

The efficiency is calculated as:

$$\text{efficiency} = \frac{\text{energy output}}{\text{energy input}}$$

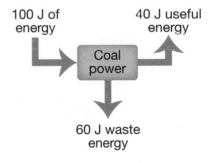

100 J of energy

40 J useful energy

Coal power

60 J waste energy

a   Calculate the efficiency of the coal-fired power station.

b   How does your answer compare to the value in the table?

Solar power

Biomass

Hydroelectric power

Task 2

At the end of 2006, the world coal reserves were about 900 billion tonnes ($900\,000\,000\,000$ or 9×10^{11} tonnes). In 2007, worldwide, 7.075 billion tonnes of coal were extracted.

1   Based on the data for 2006–07, calculate the number of years before coal supplies run out, if the rate of consumption stays the same.

2   The annual increase in coal consumption is around 2.5%. Compound growth rates can be calculated from the formula below, where N is the amount of coal extracted in the current year, N_0 is the amount of coal extracted in 2006, r is the increase per year of 2.5% and n is the number of years since 2006.

$$N = N_0 \left(1 + \frac{r}{100}\right)^n$$

   a How much coal would you expect the world to use in the year 2020?
      (In this case, n is the difference between 2020 and 2006.)

   b Using your value from part **a**, calculate how long the world's 2006 coal reserves would last at the 2020 rate of extraction.

   c In pairs, discuss why your new estimate might not be valid.

   d Investigate what would happen if the rate of usage increases further.

Task 3

Many European countries are trying to reduce their carbon footprint, and conserve their non-renewable energy reserves, by switching to renewable sources such as wind, solar and tidal.

The map shows percentages, p, of energy produced by renewable sources in European countries.

1   Use the data sheet to find the names of the countries in the map.

   a Use an appropriate method to represent the information.

   b Draw a pie chart to show the number of countries in each category, from $p < 5\%$ to $p \geqslant 40\%$.
      Label your pie chart and give it a suitable title.

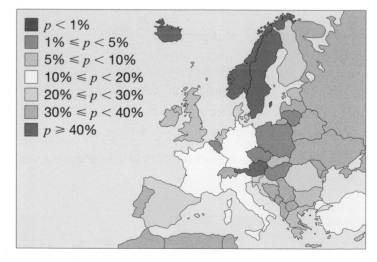

$p < 1\%$
$1\% \leqslant p < 5\%$
$5\% \leqslant p < 10\%$
$10\% \leqslant p < 20\%$
$20\% \leqslant p < 30\%$
$30\% \leqslant p < 40\%$
$p \geqslant 40\%$

2   Write a short report, no more than 250 words long, explaining clearly what your pie chart shows and how this represents the information better than the map does.

Task 4

Non-renewable energy sources are running out. Alternatives, including renewable resources, are increasingly being used.

1   Work in a small group. Study the two pie charts and then write statements comparing electricity generation in 1971 and 2004.

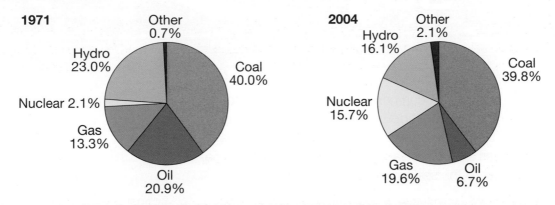

1971
Other 0.7%
Hydro 23.0%
Coal 40.0%
Nuclear 2.1%
Gas 13.3%
Oil 20.9%

2004
Other 2.1%
Hydro 16.1%
Coal 39.8%
Nuclear 15.7%
Gas 19.6%
Oil 6.7%

Generation of energy worldwide, by fuel (%)

Your **carbon footprint** is an estimate of how much carbon dioxide is produced to support your lifestyle.

Factors that contribute to your carbon footprint include how much electricity you use at home, whether you travel by car or public transport and even the carbon dioxide emitted to produce the food you eat and the clothes you wear.

2   Use the internet to find an online carbon footprint calculator. Answer the questions on the website and find your own carbon footprint.

Task 5 (extension)

Use internet searches, together with the information you have found in this activity, to summarise ways of reducing a typical carbon footprint, both individually and in the home. Investigate and compare measures such as switching to energy-efficient lightbulbs, fitting loft insulation or turning off appliances when they are not in use, and at least three other suggestions.

Produce a poster, presentation or documentary to encourage people to reduce their carbon emissions. Include mathematical diagrams and charts to support your arguments.

HOW DID YOU FIND THESE TASKS?

- What did you find easy or difficult about these tasks?
- Did you work on your own, in pairs or in groups, and how did this help or hinder your approach and success with these tasks?
- What did you learn about how maths is used and applied in real-world situations?

Can we hold back the sea?

As a small island, the UK can hardly afford to lose land to the sea. Water, one of the most powerful forces of nature, has been wearing away the British coastline for centuries. As fast as coastal defences are put in place at one part of the coastline, more land is lost to the sea further down the coast.

Task 1

The UK coastline is about 11 073 miles long. The total area of the UK is close to 94 600 square miles. 1 mile is equivalent to 1600 m.

a Investigate the area of rectangles with perimeter of 20 cm. How many rectangles can you find with sides that are whole numbers of centimetres? Comment on your answers.

b If the average rate of coastal erosion is 88 cm per year, how long might it take the sea to consume the United Kingdom? Comment on your answer.

Certain stretches of the UK coastline are particularly vulnerable to coastal erosion. One such area is Spurn Head, part of the Holderness coast.

In such areas, coastal erosion is carefully monitored. Since erosion is a slow process, monitoring is done over many years.

Coastal erosion can be measured and mapped using a graph. As erosion varies, future erosion is predicted using average values.

Task 2

A conservation group has been monitoring a 1 kilometre stretch of coastline for the past 50 years (1960–2010). Location posts were positioned along a straight baseline and the distance from each post to the coastline, perpendicular to the baseline, was measured every year.

The table on data sheet 1 shows the current distance of each location point from the coastline, the height of the cliff above sea level, the total eroded since 1960 and the average erosion per year.

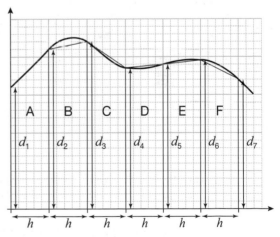

1   Draw a graph to investigate whether there is any correlation between the height of the cliff and the total erosion over the last 50 years.

2   Draw a graph to show the current position of the coastline and its position in 1960.

You can estimate the area under a graph using the trapezium rule.

Area under the graph

= area A + area B + ... + area F.

$$= \tfrac{1}{2}(d_1 + d_2)h + \tfrac{1}{2}(d_2 + d_3)h + ... + \tfrac{1}{2}(d_6 + d_7)h$$

This can be rewritten as:

$$\left(\frac{d_1}{2} + \frac{d_2}{2}\right)h + \left(\frac{d_2}{2} + \frac{d_3}{2}\right)h + ... + \left(\frac{d_6}{2} + \frac{d_7}{2}\right)h$$

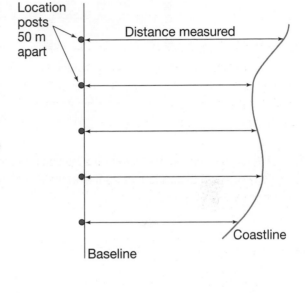

Because the trapeziums are all of equal width, h is a common factor, so this gives:

$$h\left(\frac{d_1}{2} + \frac{d_2}{2} + \frac{d_2}{2} + \frac{d_3}{2} + \frac{d_3}{2} + \frac{d_4}{2} + \frac{d_4}{2} + \frac{d_5}{2} + \frac{d_5}{2} + \frac{d_6}{2} + \frac{d_6}{2} + \frac{d_7}{2}\right)$$

$$= h\left(\frac{d_1}{2} + d_2 + d_3 + d_4 + d_5 + d_6 + \frac{d_7}{2}\right)$$

This is the trapezium rule.

3   Use the trapezium rule to find the amount of land lost between 1960 and 2010.

Task 3

Coastal regions have been hit hard by flooding in recent years, as the rainwater headed down the rivers into the sea.

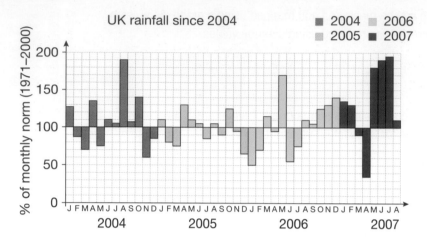

1   Use the information in this bar chart to comment on the rainfall patterns for the period from 2004 to 2007. Consider both the pattern of rainfall and the annual rainfall.

In June 2007, the UK suffered extensive flooding nationwide.

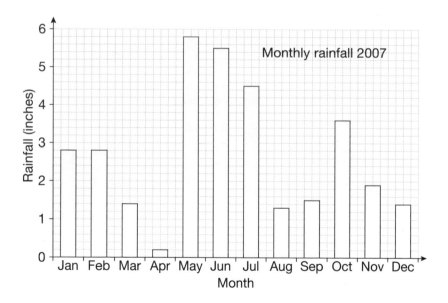

2   Find the average rainfall, in millimetres, for 2007.

3   Based on the information in both graphs, how would you expect this weather pattern to affect coastal erosion?

Task 4

The rate of the erosion of the coast can be slowed down, or even stopped, by expensive coastal engineering. In extreme cases, land may be reclaimed. Holland, in the Netherlands, has reclaimed a lot of the land that it now farms and builds on.

1   The map of Holland, on data sheet 2, shows land that has been reclaimed from the sea. Find the approximate area of Holland, including the reclaimed land.

2   Now find the approximate area of land that has been reclaimed.

3   It is estimated that 70% of Holland is below sea level. Is this statement correct?

Task 5 (extension)

Coastal erosion and rising sea levels are world-wide problems. As sea levels rise, coastline defences and methods of reclaiming land from the sea become more important.

Investigate the problems and produce a presentation explaining the threat of land loss to the sea and the methods and materials used to prevent it or to reclaim land.

You could consider:

- whether coastal erosion is linked to rising sea levels

- the rate at which the sea level is rising and how long it will be before various towns and cities in the UK are below sea level

- how the UK is improving its coastline defences and how these methods compare to those used by other countries

- how countries such as Holland and Dubai are exploiting these methods to reclaim land which is already below sea-level.

HOW DID YOU FIND THESE TASKS?

- What did you find easy or difficult about these tasks?
- Did you work on your own, in pairs or in groups, and how did this help or hinder your approach and success with these tasks?
- What did you learn about how maths is used and applied in real-world situations?

Facebook

Facebook is a social networking website. Friends use it to connect to each other, and then send messages, share photos and post updates on what they are doing. It also allows people to connect with others they do not know, but who have the same interests. It is one of the biggest social networking sites in the world. In 2010 it had more than 500 million active users.

Task 1

1   Here are some facts taken from the Facebook website in 2010.

   Copy each sentence, rewriting the numbers in standard form.

   • More than 150 000 000 active users access their Facebook pages through mobile phones.

   • Over 300 000 users helped translate Facebook into other languages.

   • More than 30 000 000 000 pieces of content (such as web links, news stories, blog posts, notes, photo albums) are shared on Facebook each month.

2   Businesses can place advertisements on Facebook. They assess the success of their advertisements by measuring the **click-through rate**. This is the number of times a user clicks on their advertisement divided by the number of times the advertisement is shown.

   a   Imagine that one Facebook advertiser achieved a click-through rate of 70% for an advertisement shown 1.1 million times. How many people clicked through the advertisement? Give your answer in standard form.

   b   On a discussion forum, one business complains that, even though they had 560 000 click-throughs on their Facebook advertisement, they only achieved a click-through rate of 4%. How many times was their advertisement shown?

Task 2

Look at these diagrams.

Time spent on Facebook in June 2009 and 2010 by country (hours spent per person per month)

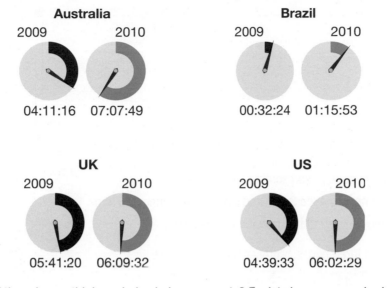

Australia
2009   2010
04:11:16   07:07:49

Brazil
2009   2010
00:32:24   01:15:53

UK
2009   2010
05:41:20   06:09:32

US
2009   2010
04:39:33   06:02:29

1   What amount of time do you think a whole circle represents? Explain how you reached your answer.

2   What is the percentage increase in time spent on Facebook from 2009 to 2010 in:

   a the US   b the UK   c Brazil   d Australia?

   Which country shows the fastest growth in time spent on Facebook? Why do you think this is?

Task 3 (extension)

Assume growth in Facebook continues at the same rate each year. In what year will Brazil reach the UK's 2010 level of hours spent per person per month?

Task 4

In these diagrams, each person is represented as a dot, called a **node**: n.

Each arrow represents a **connection** on Facebook: c.

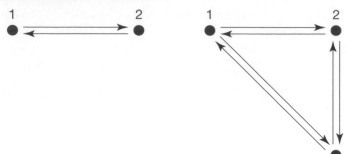

The first diagram has two nodes, representing two people.

- Node 1 can connect to node 2 → and node 2 can connect to node 1 ←.

Therefore, there are two nodes and two possible connections.

The second diagram has three nodes, representing three people.

- Node 1 can connect to node 2 → and node 3 ↘.
- Node 2 can connect to node 1 ← and node 3 ↓.
- Node 3 can connect to node 1 ↖ and node 2 ↑.

Therefore there are three nodes and six possible connections.

1 Draw the connections for four nodes and five nodes. How many possible connections are there for each of them?

2 Copy and complete this table.

n	c
2	2
3	6
4	
5	

3 Express the relationship between n and c as an equation.

4 a If a class of 30 students all connected to each other, how many connections would there be?

   b How many students would be needed to make more than 1000 possible connections?

5 Here are the numbers of people, in millions, on Facebook in Australia, Brazil, the UK and the US in 2010.

Country	People on Facebook (millions)
Australia	8.7
Brazil	28.7
UK	24.2
US	127

a Use your equation to calculate the number of possible connections on Facebook within each country. Give your answer in standard form.

b Compare your answers to part **a** for Brazil and the UK with your answers to Task 2, questions **2b** and **2c**. What would you expect to happen to the number of possible connections in each country in the future?

c Why would it not be sensible only to consider possible connections within one country?

Task 5 (extension)

Some social networking researchers claim that examining one-to-one connections between nodes only takes into account pairs of friends. It leaves out all the possible sub-groups.

1 In Task 4, the second diagram shows that there are sub-groups: (1, 2), (1, 3), (2, 3) and (1, 2, 3). Look again at your diagrams for four and five nodes, then copy and complete this table for sub-groups, s.

n	s	Sub-group list
2	1	(1, 2)
3	4	(1, 2), (1, 3), (2, 3), (1, 2, 3)
4		
5		

2 Express the relationship between n and s as an equation.

3 If a class of 30 students all connected to each other, how many sub-groups would there be?

4 Consider your answer to question **3** and your answer to Task 4, question **4a**. What do you think of these different ways of analysing possible connections on Facebook?

Task 6

Imagine the marketing department of a company would like to create a Facebook page for customers who like their products, but the senior management team is undecided. With what you have learnt about Facebook, use mathematics as a tool to write a presentation to persuade the team. (You may collect other statistics and information about Facebook from the internet.)

HOW DID YOU FIND THESE TASKS?

- What did you find easy or difficult about these tasks?
- Did you work on your own, in pairs or in groups, and how did this help or hinder your approach and success with these tasks?
- What did you learn about how maths is used and applied in real-world situations?

Population and pensions

A pension is probably the furthest thing from your mind right now, but as people live longer, the short-fall in the pension pot of money increases and the age before you can retire increases. A private pension is definitely something to consider.

When you start work you will pay **national insurance**, which will give you a **state pension**, but can you afford to live on what the Government will give you when you stop working? For more and more retired people, the answer to this question is: 'No!'

In October 2010, the state pension for a single person was £97.65 per week and for a married couple it was £156.15.

Task 1

The population of the UK is changing. In 2008, population growth was about 0.7%.

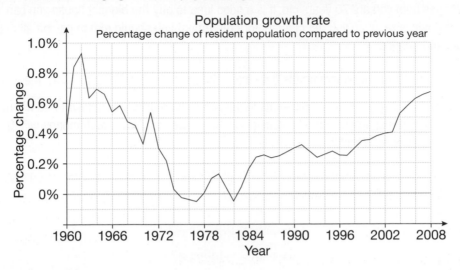

Population growth rate
Percentage change of resident population compared to previous year

1   **a** Study the line graph and explain its key features.

    b In 2008, the estimated population of the UK was 60 944 000. Estimate the population in 2009.

    c The census in 2001 showed the UK had a population of 58 789 194. If the population growth was, on average, 0.5% per year, what will the population be by 2034?

2   Life expectancy in the UK is also changing. The stacked bar chart divides the population into four different age groups.

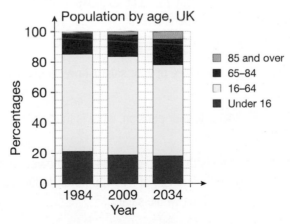

Population by age, UK

Legend:
- 85 and over
- 65–84
- 16–64
- Under 16

    a Study the bar chart and explain the predicted trend it shows.

    b Using the population figure you found in question **1b**, estimate the number of people expected to be aged 16–64 and the number of people expected to be aged 65 and over in 2034.

3   Write a short report explaining how the changing population in the UK affects the Government's 'pension pot'.

Task 2

It may seem a long time ahead but it will soon be time to plan for your retirement. Banks are now lending money for mortgages until you are 70 years old. It is a good idea to buy the biggest house you can, while you are working, then down-size when you get older. Before retirement you should aim to pay off all outstanding loans, mortgages and finance.

1 **a** Look at Kamal's bank statement on the data sheet. What are his total outgoings for the month?

   b Kamal is going to try to pay off all his debts in full before he retires. Some payments will not need to be made once he retires; others could be considered non-essential. How much money will Kamal have to pay per week after he retires?

2 **a** If, at the age of 67 years, Kamal receives the single person's retirement pension of £97.65 per week, how much money will he need to find each week?

> **Smart Bank.com**
> Investment interest
> 2.4% per year

Kamal has moved to a smaller house and put £50 000 in an investment account.

   b If Kamal lives to be 85 years old, will he always have enough money to pay his bills?

Task 3

Paying into a private pension is a very good idea, but you need to make a commitment not to touch the money until you are at least 55 years old. Most company pensions require a contribution from the employee but also a similar or higher percentage from the employer as well.

1 **a** If you earn £19 400 a year and contribute 6% of your salary, how much will you pay into your pension each year?

   b If your employer contributes 8% towards your pension, what is the total amount contributed to your pension per year?

Pension contributions are not taxed. Someone with a tax code of 647L is allowed to earn £6470 before they start paying tax. The basic tax rate is 20% on their salary of £19 400 (after deduction of their tax allowance). National insurance contribution is set at 11% of the whole salary.

2 Investigate the financial implications of paying 6% of your salary into a pensions scheme.

Task 4

The Government needs to find ways to make money to cover the short-fall in pensions. The first thing they did was to raise the retirement age for men from 65 to 68 and for women from 60 to 68 by 2046. This figure may continue to rise, perhaps to 70 in coming years.

Population: by gender and age, mid 2009

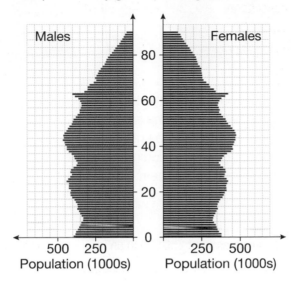

1   Use this population pyramid to approximate the number of men and woman aged between 65 and 70 in mid-2009.

2   In 2009, the average salary in the UK was reported to be £28 207. Use this figure to estimate the additional revenue the government could get from increasing the retirement age from 65 to 70.

3   Comment on your result and give reasons why the actual value could be much lower.

Task 5 (extension)

There are many different types of pension available. When choosing a pension, one important thing to consider is whether payments are based on final salary or average salary.

Use the internet to research the different types of pension, and other ways of saving, available.
Produce a leaflet to inform people about retirement and their options.

HOW DID YOU FIND THESE TASKS?

- What did you find easy or difficult about these tasks?
- Did you work on your own, in pairs or in groups, and how did this help or hinder your approach and success with these tasks?
- What did you learn about how maths is used and applied in real-world situations?

London black cabs

The famous London black cabs provide one of the best taxi services in the world. That's because all drivers have to pass a special test called 'the knowledge'. They have to know 320 of the shortest possible routes across London, including lots of tourist sites. It takes two to four years to learn the routes.

Task 1

1 The velocity–time graph shows the first 20 seconds of a London black cab's journey in the rush hour.

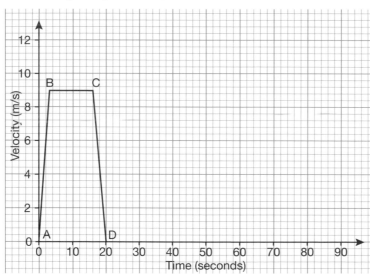

The cab starts at rest, having dropped off a passenger at a theatre. It accelerates at a constant rate over 3 seconds to 9 m/s (AB) and then travels at a constant velocity for 13 seconds (BC) before coming to rest again in a further 4 seconds (CD), when it stops at a zebra crossing.

Find:

a the acceleration over the first 3 seconds

b the acceleration over the next 13 seconds

c the acceleration over the last 4 seconds.

2 Copy the graph and complete it to show the cab's journey until it picks up the next passenger.

- The cab stops at the zebra crossing for 7 seconds while a pedestrian crosses the road.

- It accelerates at a constant rate over 4 seconds to 11 m/s.

- It travels at constant velocity for 9 seconds.

- It comes to rest again in 5 seconds, behind a stationary bus.

- It rests there for 6 seconds before moving on.

- It accelerates at a constant rate over 2 seconds to 6 m/s.

- It travels at a constant velocity for 4 seconds.

- The road becomes clear ahead, so it accelerates at a constant rate again over 2 seconds, to 12 m/s.

- It travels at a constant velocity for 19 seconds.

- It decelerates at a constant rate over 3 seconds, to 4 m/s, as the driver sees a potential passenger.

- It comes to rest in 9 seconds to pick up the passenger.

3 Use your completed graph to find the taxi's acceleration:

a over 4 seconds, pulling away from the zebra crossing

b over 5 seconds, waiting behind a bus

c over 2 seconds, moving again behind the bus

d over 2 seconds, the road being clear ahead

e over 3 seconds, having seen the waiting passenger

f to rest to pick up the passenger.

4 This velocity–time graph shows a different black cab's rush-hour journey. Compare this journey to your velocity–time graph for the first cab driver in terms of:

a the time spent between dropping off one passenger and picking up another

b the number of stops and starts

c the maximum velocities reached

d the acceleration achieved.

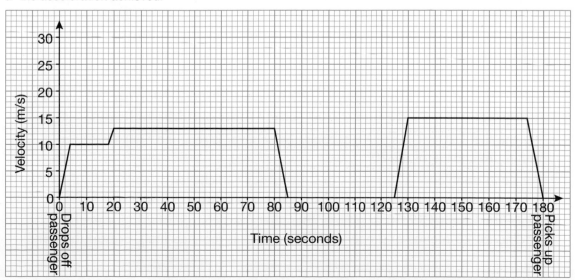

Task 2 (extension)

This second cab driver was driving through a zone where the speed limit was 30 miles per hour. Comment on his velocity.

Task 3

You can calculate the distance travelled by finding the area under the velocity–time graph.

1   Look at the graph you used in question **1** of Task 1. Calculate the distance travelled between dropping off the passenger at the theatre and stopping at the zebra crossing.

2   Look at the graph you drew for question **2** of Task 1. Calculate the total distance travelled by the cab, from the time it dropped off its passenger to the time it picked up another passenger.

3   Compare the distance travelled between dropping off and picking up passengers for the first cab with the distance travelled between dropping off and picking up passengers for the second cab.

Task 4

London black cab fares are calculated approximately as follows.

Tariff 1: Monday–Friday between 06:00 and 20:00 (other than a public holiday)

- For the first 280 metres or 60 seconds (whichever is reached first) there is a minimum charge of £2.20
- For each additional 140 metres or 30 seconds (whichever is reached first), or part thereof, if the fare is less than £15.80 then there is a charge of 20p
- Once the fare is £15.80 or greater then there is a charge of 20p for each additional 98 metres or 21 seconds (whichever is reached first), or part thereof

Tariff 2: Monday–Friday between 20:00 and 22:00 or during Saturday or Sunday between 06:00 and 22:00 (other than a public holiday)

- For the first 220 metres or 48 seconds (whichever is reached first) there is a minimum charge of £2.20
- For each additional 120 metres or 24 seconds (whichever is reached first), or part thereof, if the fare is less than £19.00 there is a charge of 20p
- Once the fare is £19.00 or greater then there is a charge of 20p for each additional 98 metres or 21 seconds (whichever is reached first), or part thereof

Tariff 3: For any hiring between 22:00 on any day and 06:00 the following day or at any time on a public holiday

- For the first 180 metres or 36 seconds (whichever is reached first) there is a minimum charge of £2.20

- For each additional 80 metres or 18 seconds (whichever is reached first), or part thereof, if the fare is less than £23.00 there is a charge of 20p

- Once the fare is £23.00 or greater then there is a charge of 20p for each additional 98 metres or 21 seconds (whichever is reached first)

1 This graph shows a short passenger journey made last Tuesday at 7.20 pm in a London black cab. What was the fare?

2 Compare this fare with what the journey would have cost if it had been 1 hour later.

3 Compare this fare with what the journey would have cost if it had been 4 hours later.

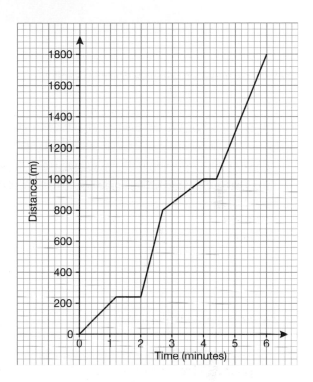

Task 5 (extension)

Imagine you are the Mayor of London. You have received a request to allow more cabs on London's roads. Meanwhile, an environmental pressure group would like you to reduce the number of London black cabs. You must consider the social, environmental and economic implications. Write a request to your advisor, spelling out the kinds of mathematical information you require to help you make your decision.

HOW DID YOU FIND THESE TASKS?

- What did you find easy or difficult about these tasks?
- Did you work on your own, in pairs or in groups, and how did this help or hinder your approach and success with these tasks?
- What did you learn about how maths is used and applied in real-world situations?

Extreme sports

Many popular sports involve speed. Extreme sports, including **zorbing** and **skydiving**, also have high levels of danger associated with them.

Zorbing, also known as sphereing or orbing, is a relatively new extreme sport. A zorb is a large inflatable sphere, with a smaller ball inside. The two are connected by means of about 1000 nylon ropes. The participant climbs into the inner ball and rolls down a slope.

Skydiving is a more established sport. Skydivers jump out of an aeroplane and descend for quite a distance, in freefall, before opening a parachute.

Task 1

The inflatable outer ball of a zorb has a diameter of 3 m. The smaller inner sphere has a diameter of 2 m.

The **volume**, V, of a sphere is given by the formula: $V = \frac{4}{3}\pi r^3$

and the **surface area**, A, is given by the formula: $A = 4\pi r^2$ where r is the radius.

1 What is the total area of plastic needed to make both the inner and outer balls?

2 What is the volume of air inside the inner ball?

3 What is the volume of the air cushion between the inner and outer balls?

Task 2

Runs are usually 200 m long, although some are 300 m long. A zorb can reach a speed of 50 kilometres per hour (km/h).

How quickly do you think the zorb rotates? How long would a zorb run take? This task will help you find out.

1 What is the circumference of the outer ball?

2 **a** How many times will the ball rotate on a run of 200 metres?

    b How many times will the ball rotate on a run of 300 metres?

3 Copy and complete the table to show how long a zorb run would take at different speeds.

Length of run	10 km/h	20 km/h	30 km/h	40 km/h	50 km/h
200 m					
300 m					

4 A typical price for zorbing is £30 per run. How much is that per second?
Use your results from question **3** to copy and complete this table.

Length of run	10 km/h	20 km/h	30 km/h	40 km/h	50 km/h
200 m					
300 m					

Task 3

If you were a skydiver, as soon as you jumped out of an aeroplane you would start to fall, faster and faster, pulled down by **gravity**.

You would not keep falling faster all the way down, though, as **air resistance** cancels out gravity's accelerating force. You would reach a constant speed, called a **terminal velocity**.

You would carry on at that speed until you opened your parachute, when your speed would immediately and suddenly change to a lower constant velocity.

1 Draw a sketch graph to show a skydiver's descent. Show height on the vertical axis and time on the horizontal axis. Your graph should show three stages: the **initial acceleration**, the **terminal velocity** and the **velocity after the parachute opens**.

Now suppose a skydiver jumps from an aeroplane at an altitude of 10 000 metres. Another skydiver follows him, 30 seconds later.

You can model their falls with graphs.

The first skydiver's altitude is given by the equation $d = 10\,000 - 5t^2$, where d is his **altitude** (his distance above ground), in metres, and t is the time, in seconds, since he left the aeroplane. He follows this equation for the first seven seconds, after which he reaches terminal velocity.

After seven seconds, he falls at constant velocity, given by the equation $d = 10\,280 - 75t$.

A minute after jumping, he opens his parachute and slows down immediately. His altitude is given by $d = 6200 - 7t$.

2 Show this information on a graph.

The second skydiver is going to open her parachute when she catches up with the first. She jumps when $t = 30$ (30 seconds after the first skydiver jumped).

Her equation while accelerating is given by $d = 5500 + 300t - 5t^2$.

Seven seconds later, she reaches terminal velocity, and her equation becomes $d = 12\,530 - 75t$.

3 Show this information on the same set of axes, and use the graphs to find the time at which they will be at the same altitude. At this point the second skydiver opens her parachute.

Task 4

Your first try at skydiving will cost about £155. You will jump from a height of 4000 metres.

- For the first seven seconds, your altitude is given by the equation $d = 4000 - 5t^2$, until you reach terminal velocity of 75 metres per second (m/s).

- You open the parachute at a height of 1650 m.

- You then continue towards the ground at a speed of 7 m/s.

1 Calculate the total time for the whole jump, and the cost per second.

2 Which is cheaper, per second, zorbing or skydiving?

Task 5 (extension)

Research any world records that relate to skydiving. How do the records compare to the figures you have used in Task 3?

HOW DID YOU FIND THESE TASKS?

- What did you find easy or difficult about these tasks?
- Did you work on your own, in pairs or in groups, and how did this help or hinder your approach and success with these tasks?
- What did you learn about how maths is used and applied in real-world situations?

Maths – music to your ears?

The idea that there are links between maths and music is not new. Pythagoras, the Greek mathematician, acknowledged such links many centuries ago.

As he walked past a blacksmith's workshop, Pythagoras thought that the sounds made by the hammers on the metal were in **harmony**. On studying the hammers he found that the masses of those that produced these harmonies were all in simple ratios such as 2 : 1 or 4 : 1. The masses of hammers producing non-harmonious sounds were not in such simple ratios.

During the seventeenth century, the German mathematician Gottfried Leibniz wrote to a friend about musical theory. To justify his musical reasoning, he used **Euclid's algorithm**. This has its origins back in 325–265 BC.

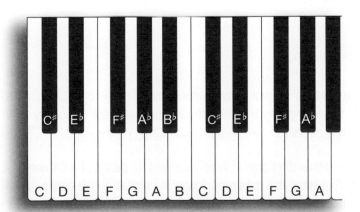

Task 1

Musical notes are arranged into groups called **octaves**, each made up of 13 notes. Every note is one semitone higher than the one before it in the octave. Two semitones make a whole **tone**. The notes are identified by the letters A, B, C, D, E, F and G. Some notes are called sharp (♯) or flat (♭).

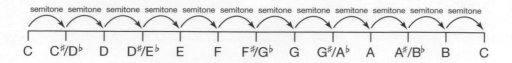

Western music is often based on an 8-note scale taking eight notes from the 13 described above. For example, the C major scale has the following pattern.

C   D   E   F   G   A   B   C

1   Write down the number of semitones between each pair of adjacent notes in the scale of C major. Remember that two semitones make one tone.

2   All the major scales follow the same pattern of tones and semitones as the scale of C major. Write out the sequence of notes in the scale of D major, starting at D and following the pattern you have just worked out.

3   Draw up a table that shows the notes for the major scales of C, D, E, F, G, A and B.

Task 2

Refer to data sheet 1 for this task.

1   Use the table of notes and frequencies to work out the frequency of the notes A2 and A5, which are an octave below, and two octaves above, the note A3.

2   On the same set of axes, plot the frequencies of the notes in the major scales of C, D and E. Comment on the shape of your graphs.

Task 3

The frequency of a note is inversely related to the time interval between consecutive peaks of a sound wave reaching our ears. You can calculate the time between peaks by taking the **reciprocal** of the frequency. This time is called the **period** and is measured in seconds.

1   Rachel hears a note of C4. Draw a table to show the times at which the next 10 peaks of a C4 note reach her ear. Do the same for C5. Draw a chart to show this data graphically. How often do the peaks for C4 and C5 line up? How can you explain this?

2   How many peaks would a note of C5 produce in the time a C4 produced 10? Express this as a ratio, in its simplest terms.

Task 4

A chord is produced when two or more notes are played together. Some chords sound better than others. There are rules that govern which notes sound good when played together.

1   In your chart from question **1c** of Task 1, highlight the columns marked **3rd** and **5th**. What are the third and fifth notes in the scale of C major?

2   Draw a graph to show the times at which the peaks of a C4 note, and its major 3rd and 5th notes, arrive at a listener's ear. Plot all the peaks that would arrive within 0.5 seconds.

3   Write down the number of peaks a C4 note produces by the time it lines up with its 3rd note. How many peaks does the 3rd note produce in this time? Express this as a ratio in its simplest terms.

4   Write down the number of peaks a C4 note produces by the time it lines up with its 5th note. How many peaks does the 5th note produce in this time? Express this as a ratio in its simplest terms.

5   How often do all three sets of peaks line up? What is the relationship to the ratios you have calculated?

Task 5 (extension)

The strings on a guitar or ukulele are all the same length. When a string is plucked, allowing it to vibrate freely along its whole length, it produces a note that is called a **fundamental**. The metal bars along the neck of a guitar or ukulele are called **frets**. When the string is pressed against a fret it is shortened. This produces a higher note when the string is plucked. The frets are positioned to produce notes a semitone apart.

1   You are working for a ukulele maker who has asked you to put the frets in the correct position on the neck. The ukulele has strings 32 cm long. The length of the string is inversely proportional to the frequency of the note produced. Where on the neck would you place the fret to produce a note one octave higher than the fundamental? Write your answer in the table on data sheet 2.

2   One of the strings on a ukulele is tuned to the fundamental note of C4. Where on the neck would you place the fret to produce the third note and the fifth note in the scale of C major? Use your answer from Task 4 to help you. Write your answers in the table on the data sheet.

3   Use the formula on the data sheet to work out where to place the remaining nine frets. Fill in the table.

4   Draw the frets onto the diagram on the data sheet. Remember that all your results are measured from the bridge of the ukulele. What do you notice about the spacing of the frets along the length of the ukulele's neck?

HOW DID YOU FIND THESE TASKS?

- What did you find easy or difficult about these tasks?
- Did you work on your own, in pairs or in groups, and how did this help or hinder your approach and success with these tasks?
- What did you learn about how maths is used and applied in real-world situations?

Acknowledgements

The publishers wish to thank the following for permission to reproduce photographs, statistics and illustrations. Every effort has been made to trace copyright holders and to obtain their permission for the use of copyright material. The publishers will gladly receive any information enabling them to rectify any error or omission at the first opportunity.

Photographs:

p.6 © MARKBZ/shutterstock.com, © Sebastian Knight/shutterstock.com, © Luca di Filippo/shutterstock.com; p.7 © Sebastian Knight/shutterstock.com, © Sharon Howe/shutterstock.com, © gregg williams/shutterstock.com, © Martin Pateman/shutterstock.com, © Francis Bossé/shutterstock.com, © Peter zijlstra/shutterstock.com, © Vishnevskiy Vasily/shutterstock.com; p.10 © EDHAR/shutterstock.com; p.11 © Yuri Arcurs/shutterstock.com; p.13 © r.nagy/shutterstock.com; p.14 © Sebastian Kaulitzki/shutterstock.com; p.15 © David H. Lewis/iStockphoto.com; p.16 © Charlie Hutton/shutterstock.com; p.18 © Tobias Keckel/iStockphoto.com, © Janez Habjanic/shutterstock.com, © jannoon028/shutterstock.com, © Ensuper/shutterstock.com, © Viktor Gmyria/shutterstock.com; p.21 © Coprid/shutterstock.com; p.22 © Andrew Howard/shutterstock.com, © H.Brauer/shutterstock.com; © KA Pike/biodieselpictures.com; p.23 © Moreno Soppelsa/shutterstock.com; p.24 © Gordon Milic/shutterstock.com; p.26 © Ricardo Cervera/RexFeatures.com; p.30 © Nikolay Stefanov Dimitrov/shutterstock.com, © nicoolay/iStockphoto.com; p.31 © Pichugin Dmitry/shutterstock.com; p.34 © Alta Oosthuizen/shutterstock.com, © Semjonow Juri/shutterstock.com, © André Gonçalves/shutterstock.com; p.36 © Kirsanov/shutterstock.com, © LilKar/shutterstock.com; p.37 © Peter Waters/shutterstock.com; p.38 © Bianda Ahmad Hisham/shutterstock.com, © rm/shutterstock.com; pp.38-39 © Brendan Howard/shutterstock.com; p.40 © Jaap2/iStockphoto.com, © Ruvan Boshoff/iStockphoto.com, © Brendan Howard/shutterstock.com; pp.40-41 © granata1111/shutterstock.com; p.42 © Sven Hoppe/shutterstock.com; p.48 © michele lugaresi/iStockPhoto.com, © r.nagy/shutterstock.com, © Stephen Finn/shutterstock.com; p.50 © Keith Gentry/shutterstock.com; p.52 © Vincente Barcelo Varona/shutterstock.com; p.55 © Victor Melniciuc/iStockphoto.com, © A.S. Zain/shutterstock.com; p.58 © Davide69/shutterstock.com; p.62 © Jess Wiberg/iStockphoto.com, © Joseph Nicephore/commons.wikimedia.org, © basel101658/shutterstock.com; p.64 © Nikuwka/shutterstock.com; p.65 © Joern/shutterstock.com, © Tallllly/shutterstock.com, © Dmitriy Shironosov/shutterstock.com, © Ivica Drusany/shutterstock.com, © Vladitto/shutterstock.com; p.66 © Sze Fei Wong/iStockphoto.com, © Serge Bikhunenko/shutterstock.com, © Grekoff/shutterstock.com, © Angela Luchianiuc/shutterstock.com, © Faraways/shutterstock.com; p.67 © Tatiana Popova/shutterstock.com, © Dawid Zagorski/shutterstock.com; p.68 © vladimirs guculaks/iStockphoto.com; p.70 © Andriano/shutterstock.com, © Worldpics/shutterstock.com; p.72 © Nitipong Ballapavanich/shutterstock.com; p.73 © lidian/shutterstock.com, © Becky Stares/shutterstock.com; p.76 © olly/shutterstock.com, © Tim Jenner/shutterstock.com; p.79 © Viktor Gmyria/shutterstock.com; p.80 © Kuzma/shutterstock.com, © John Tenniel; p.83 © Valentyn Volkov/iStockphoto.com; p.86 © Tom Bilek/shutterstock.com; p.90 © Jane Rix/shutterstock.com; p.94 © Eastimages/shutterstock.com; p.96 © Utekhina Anna/shutterstock.com, © arbit/shutterstock.com; p.97 © firek1/shutterstock.com; p.98 © ilolab/shutterstock.com; p.99 © yellowj/shutterstock.com, © Steve Mann/shutterstock.com; p.102 © Scott Cramer/iStockphoto.com, © oneclearvision/shutterstock.com; p.103 © rozdesign/ iStockphoto.com; p.104 © Robert Dant/ iStockphoto.com; p.106 © Krom/shutterstock.com, © JazzBoo/shutterstock.com; p.107 © l i g h t p o e t/shutterstock.com; p.109 © 3dimentii/shutterstock.com; p.110 © Karin Hildebrand Lau/shutterstock.com, © Monkey Business Images/shutterstock.com; p.111 © Jacqueline Abromeit/shutterstock.com; p.112 © Serghei Starus/shutterstock.com; p.114-7 courtesy of NASA; p.118 © YanLev/shutterstock.com, © Duncan Walker/iStockphoto.com; p.119 © D. Roberts/Science Photo Library, © Becky Stares/shutterstock.com; p.120 © D Barton/shutterstock.com; p.122 © empipe/shutterstock.com; p.123 © Monkey Business Images/shutterstock.com, © niderlander/shutterstock.com, © JinYoung Lee/shutterstock.com; p.124 © koksharov Dmitry/iStockphoto.com; p.127 © Mike Dabell/iStockphoto.com; p.130 © gosn.Momcilo/shutterstock.com, © Ashley Pickering/shutterstock.com; p.133 © Hans Fredrik Svartdahl/iStockphoto.com; p.134 © TebNad/iStockphoto.com, © Bullwinkle/shutterstock.com, © majeczka/shutterstock.com, © Vojtech Soukup/iStockphoto.com; p.135 © nostal6ie/shutterstock.com, © Igor Grochev/shutterstock.com, © Carlo Taccari/shutterstock.com; p.138 © Martin Kemp/iStockphoto.com, © Worakit Sirijinda/shutterstock.com; p.142 © Alberto Zornetta/shutterstock.com; p.143 © ARENA Creative/shutterstock.com; p.146 © GSPhotography/shutterstock.com, © Don McGillis/iStockphoto.com, © susandaniels/ iStockphoto.com; p.150 © Monkey Business Images/shutterstock.com; p.154 © mountainberryphoto/iStockphoto.com, © Vetlugin Evgeny Aleksandrovich/shutterstock.com; p.156 © Goran Cakmazovic/shutterstock.com; p.157 © RoxyFer/shutterstock.com, © c./shutterstock.com.

Statistics and illustrations:

pp.10-11 © Crown/www.statistics.gov.uk; p.20 © RAJAR/Ipsos MORI – MIDAS October 2008/www.rajar.co.uk; p.25 © SRI Consulting Biodiesel Report 2008/sriconsulting.com; p.35 © British Beekeeping Association/britishbee.org.uk; p.38 © Crown/royalmint.com; p.48-49 © Crown/www.statistics.gov.uk; p.59 © Crown/www.mcga.gov.uk; p.66 © United Kingdom Tea Council/www.tea.co.uk; p.75 © Westmoreland Gazette/thewestmorlandgazette.co.uk; © CollinsBartholemew; p.84 © ISAAA/isaaa.org; p.85 © Deloitte/deloitte.co.uk; pp.90-1 © ICES/ices.dk; pp. 92-3 © MFA/MMO/marinemanagement.org.uk; p.95 © BSCS. Used with permission. "Sleep, sleep disorders, and biological rhythms, BSCS and NIH", © Science, 152: 3722, "Ontogenetic Development of the Human Sleep-Dream Cycle," pp.604-619/sciencemag.org; p.98 © Crown/UK Statistics Authority/statisticsauthority.gov.uk; p.108 © Crown/direct.gov.uk; p.113 © The Sun/thesun.co.uk; p.127 © Crown/statistics.gov.uk; p.132 © Crown/statistics.gov.uk; p.136 © U.S. Energy Information Administration 2007/www.eia.doe.gov; p.139 © CollinsBartholemew; p.140 © Andrew Leaper 2007/www.zen40267.zen.co.uk; pp.142-3 © The Nielsen Company/uk.nielsen.com; p.147 © Crown/statistics.gov.uk; p.149 © Crown/statistics.gov.uk; p.152 © Crown/tfl.gov.uk.